Design and Analysis of Accelerated Tests for Mission Critical Reliability

Michael J. LuValle
Bruce G. Lefevre
SriRaman Kannan

CHAPMAN & HALL/CRC

A CRC Press Company
Boca Raton   London   New York   Washington, D.C.

Library of Congress Cataloging-in-Publication Data

LuValle, Michael.
    Design and analysis of accelerated tests for mission critical reliability / by Michael LuValle, Bruce G. Lefevre, SriRaman Kannan.
        p. cm.
    Includes bibliographical references and index.
    ISBN 1-58488-471-1 (alk. paper)
    1. Accelerated life testing. 2. Reliability (Engineering) I. Lefevre, Bruce G. II. Kannan, SriRaman. III. Title.

TA169.3.L88 2004
620'.00452--dc22

2003069580

This book contains information obtained from authentic and highly regarded sources. Reprinted material is quoted with permission, and sources are indicated. A wide variety of references are listed. Reasonable efforts have been made to publish reliable data and information, but the author and the publisher cannot assume responsibility for the validity of all materials or for the consequences of their use.

Neither this book nor any part may be reproduced or transmitted in any form or by any means, electronic or mechanical, including photocopying, microfilming, and recording, or by any information storage or retrieval system, without prior permission in writing from the publisher.

The consent of CRC Press LLC does not extend to copying for general distribution, for promotion, for creating new works, or for resale. Specific permission must be obtained in writing from CRC Press LLC for such copying.

Direct all inquiries to CRC Press LLC, 2000 N.W. Corporate Blvd., Boca Raton, Florida 33431.

Trademark Notice: Product or corporate names may be trademarks or registered trademarks, and are used only for identification and explanation, without intent to infringe.

Visit the CRC Press Web site at www.crcpress.com

© 2004 by CRC Press LLC

No claim to original U.S. Government works
International Standard Book Number 1-58488-471-1
Library of Congress Card Number 2003069580
Printed in the United States of America 1 2 3 4 5 6 7 8 9 0
Printed on acid-free paper

Preface

In the early days of accelerated testing, the method was touted as a way to compress years of use into days of testing. The practitioner subjected the device to "higher" stress than it would see in operation, and time was compressed in a nice, predictable fashion, with the parameters the only thing left to be determined. Most statistical theory for accelerated testing is built around this kind of assumption.

Although the simplest physical model does have this form, the simplest physical model does not always hold. Thus, we need to understand what the more complex physical models may look like and how to model them. Further, with the direction certain branches of engineering are taking, new devices made from new materials are being put in mission-critical applications. Failure is not tolerable over a 20-year life, so what can an accelerated test say in such a situation?

The purpose of this book is to provide theory and tools necessary to attack these problems. The theory is an integration of chemical kinetics and statistics aimed particularly at designing accelerated tests and modeling the results (sudden failure, smooth degradation, and no response at all). The tools include both general approaches that can be implemented in the various computational tools available and an explicit computing environment written in Splus®.

The theory and tools we provide here are not the final word; however, they are useful, and have been used for years by one author to support mission-critical application. They are an opening salvo in an attack on the problem of extrapolation.

For the practitioner, there are several practical tools and examples provided. For the teacher and student, exercises are scattered throughout the text. For researchers, open questions abound and the software (provided as free source code in Splus) can certainly be improved.

We wish you luck and hope you find this as useful as we have.

Authors

Michael J. LuValle, Ph.D., received his doctorate from the Division of Statistics at the University of California at Davis and started his professional career as an assistant professor at Kansas State University. He moved to AT&T Bell Laboratories in 1984, which is where he started seriously addressing the problem of extrapolation of acceleration tests. He is at present a member of OFS Laboratories in the materials, processes, and reliability research group. His formal background is in mathematical statistics, but he has worked primarily at the interface of statistics and the physical sciences. He has authored or co-authored a number of publications and invited talks in this area.

Dr. LuValle, through the graces of Dr. Lefevre, has recently become involved in the development of standards for reliability.

Bruce G. Lefevre, Ph.D., received his doctorate from the University of Florida and is a former professor on the faculty of the Georgia Institute of Technology. He has served on the technical staff of Bell Laboratories, AT&T, and Lucent Technologies and is at present a consultant to OFS on fiber-optic passive components. His technical background is in materials and passive optical component design, testing, reliability, and standards. He has authored or co-authored more than 30 technical publications and articles on these subjects.

Dr. Lefevre is Chairman of IEC SC86B (Fiber Optic Interconnecting Devices and Passive Components), Co-chair of TIA FO-4.3.3 (Working Group on Reliability of Fiber Optic Interconnecting Devices and Passive Components), and a member of the U.S. Technical Advisory Groups to IEC TC86 (Fiber Optics) and SC86B. He has participated in the drafting, editing, and publication of numerous standards on design, testing, and performance of passive fiber optic components.

SriRaman Kannan, Ph.D., obtained his Bachelor of Technology from the Indian Institute of Technology (IIT–Bombay) in 1988 and Ph.D. from Rutgers University in 1994. Following a year of U.S. Department of Energy postdoctoral fellowship (1994–1995), he joined AT&T Bell Labs in 1995. Dr. Kannan has since held the positions of Member of Technical Staff and Technical Manager. At present he is a reliability specialist at the Government Communications Division of Bell Laboratories, Lucent Technologies.

Over the years, Dr. Kannan has worked on diverse issues pertaining to optical fibers, components, and devices such as thermal stability of fiber gratings, radiation effects on glasses, and hydrogen-induced effects in rare earth-doped fibers. He has developed and executed qualification plans for several state-of-the-art telecommunication components and devices, including different types of specialty fibers, gratings, fiber lasers and amplifiers, Raman resonators, and optical channel monitors.

Dr. Kannan has authored more than 25 publications and has made extensive presentations including invited talks in industry forums. He is also involved with standards work as a participant of IEC 86B, working group 5 (Reliability of Fiber Optic Interconnecting Devices and Passive Components).

Acknowledgments

To my twin boys, Brian and Connor, whose impending birth has certainly presented an immovable deadline for finishing this work, my wife Rosemarie DiDomenico, who has supported me throughout the long years of this work, Aspen LuValle for his sage counsel, and my management and colleagues who have provided encouragement and an environment in which the research presented here could be done: J.P. Mitchell, T.L. Welsher, K. Svoboda, H.M. Cohen, W. Lambert, P. Ward, L. Hines, L. Chan, D. Klinger, V.N. Nair, J. Chambers, C. Mallows, W. Joyce, F. Nash, R.L. Easton, J. Hooper, M. Carey, M. Tortorella, J. Aspell, P. Lemaire, D. Sinclair, R. Opila, R. Frankenthal, B. Eggleton, K. Walker, D. DiGiovanni, J. Abate, R. Ahrens, J. Jacques, J. LeGrange, L. Reith, L. Braun, E. Vogel, L. Copeland, and J. Mrotek, all from the Bell Labs/OFS Labs in its various incarnations; and W.Q. Meeker, J. Lawless, W. Nelson, and C.J. Wu, who have provided encouragement from outside.

Michael J. LuValle

To my wife, Sandy, who has indulged me in all manner of things including this enterprise.

Bruce G. Lefevre

To my supportive family, fellow authors, and numerous colleagues who rendered work fun over the years.

SriRaman Kannan

Contents

1. **Background** 1
   1.1 Introduction 1
   1.2 Other Approaches 2
   1.3 Foundation of Our Approach 5
   1.4 A Simple Example 6
   1.5 Organization of This Book 9
   1.6 Complement: Background Kinetics and Statistics 11
       1.6.1 Arrhenius and Relative Humidity Models 11
       1.6.2 First-Order Kinetics 14
       1.6.3 Binomial Distribution and Its Role in Reliability ... 17
       1.6.4 Inference for the Binomial Distribution 18
       1.6.5 Splus Source Code for Matrix Exponentiation 20

2. **Demarcation Mapping: Initial Design of Accelerated Tests** 25
   2.1 Analytical Theory of Thermal Demarcation Maps 27
   2.2 Designing an Acceptance Test for a Purely Thermal Process 34
   2.3 Simple Temperature/Humidity Models 37
   2.4 Designing an Acceptance Test for a Temperature/Humidity Model 38
   2.5 Mechanical Cycling Models 40
   2.6 Acceptance Testing for Mechanical Cycling Induced by Thermal Cycling 41
   2.7 Computational Demarcation Mapping 44
   2.8 Beta Binomial Interpretation of 0 Failures 50
   2.9 An Extrapolation Theorem 51
   2.10 Summary 53
   2.11 Complements to Chapter 2 53
       2.11.1 Demarcation Maps for Multiple Experiments 53
       2.11.2 Using the Freeware 55
           2.11.2.1 Thermal Demarcation Maps 56
           2.11.2.2 Temperature/Humidity Demarcation Maps 58
           2.11.2.3 Mechanical Cycling Demarcation Maps 61

3. **Interface for Building Kinetic Models** 65
   3.1 Description and Concepts behind the Interface 65
   3.2 Complement to Chapter 3: Our Interface in Splus, Kinetic Data Objects, and the GUIs to Create Them 70
       3.2.1 Creating Components of the Kinetic Model 70
           3.2.1.1 Discrete Diffusion 71

    3.2.1.2 Creep 73
    3.2.1.3 Stress Voiding 74
    3.2.1.4 One-Step Process 75
    3.2.1.5 One Step with Second Variable 77
    3.2.1.6 One Step with Stress 77
    3.2.1.7 One Step with Second Variable and Stress 78
    3.2.1.8 Glassy One Step 79
  3.2.2 Combining Submodels 81
    3.2.2.1 Competing Reactions 81
    3.2.2.2 Mixing Reactions 85
    3.2.2.3 Reversible Reactions 86
    3.2.2.4 Rejoining Reactions 87
    3.2.2.5 Sequential Reactions 89
    3.2.2.6 Simple Connection of Internal States 90
  3.2.3 Computational Demarcation Map Example
    from Chapter 2 92

4. Evanescent Process Mapping 97
 4.1 Building Blocks for the Theory 98
  4.1.1 Model Neighborhoods 98
  4.1.2 Risk Orthogonality 102
  4.1.3 Model Enumeration 102
  4.1.4 Integrating the Theory 104
 4.2 Identifying Neighborhoods of Models, Sampling, and
  "Chunking" 105
 4.3 Example 111
 4.4 Summary, Limitations of Accelerated Testing 115
 4.5 Complement to Chapter 4: Using the Evanescent Process
  Mapping Interface to Duplicate Example 4.3 116

5. Data Analysis for Failure Time Data 135
 5.1 A Simple Data Set 136
 5.2 Adding Physical Sense to the Model 146
 5.3 Analysis of a Real Data Set 148
  5.3.1 Summary 158
 5.4 Complement: Maximum Likelihood Analysis 159
  5.4.1 Law of Large Numbers 160
  5.4.2 Central Limit Theorem 160
  5.4.3 Proof of Consistency of Maximum Likelihood 161
  5.4.4 Derivation of the Distribution of the Maximum
    Likelihood Estimator 162
  5.4.5 Splus Source Code 164
 5.5 Complement: Statistical Estimation of Kinetics from
  Failure Time Data 166
 5.6 Complement: Pseudo-Maximum Likelihood Estimation 169
 5.7 Complement: The Kaplan–Meier Estimate 171

	5.8	Complement: Printed Wiring Board Data 171
	5.9	Complement: Using the Interface 175
	5.10	Complement: Exercises to Explore Some Questions in Experiment Design 190
		Problem 5.10.1 191
		Problem 5.10.2 191
		Problem 5.10.3 191
		Hints for Problem 5.10.1 191
		Hint for Problem 5.10.2 196
		Hints for Problem 5.10.3 196

6. Data Analysis for Degradation Data 197
    6.1 Motivation and Models 197
    6.2 Background for the Example 200
    6.3 Data Analysis for the Example 201
    6.4 Complement: Background Statistical Theory 206
        6.4.1 Linear Regression and Results 206
        6.4.2 Extension to Nonlinear Regression 208
        6.4.3 Extension to an Uncertain Starting Time Model 209
        6.4.4 Prediction Uncertainty and Asymptotic Relative Efficiency 212
    6.5 Complement: Using the Software to Analyze the Example Data 213
    6.6 Complement: Exercises for Data Analysis and Experiment Design 224

References 227

Appendix: Installing the Software 231

Index 233

1
Background

1.1 Introduction

Reliability is defined as the probability that a system will operate to required specifications for a predetermined operating life within a given operating environmental regime. Mission-critical reliability means the probability of failure of the system must for all practical purposes be 0. For our purposes accelerated testing is the testing of a new material system in an environmental regime different from the operating regime, in order to learn in a very short time how reliable that system will be in the operating regime for the operating life. The problems we tackle are the problems of slow changes occurring in the material system at operating conditions that can lead to failure. There is an aspect of reliability involving failure from high-stress accidents exceeding the fundamental limits of the materials (as new); we do not attempt to handle that problem here. For example, the ability of a chunk of foam to open a hole in a new space shuttle wing is not within our scope, but the processes that can weaken the wing over time so that it is more susceptible to mechanical damage are within our scope. Another aspect of reliability within our scope, handled in this book for a very specific type of problem in Chapters 2 and 4, is that of ensuring that the assumptions used in extrapolating are rigorously challenged.

Our approach to accelerated testing rests on two fundamental concepts:

1. The observable quantities that are generated from accelerated testing (and from degradation and failure of material systems in the field) arise from changes in the physical and chemical properties of the material system. Hence, by definition, the appropriate models for the processes underlying the observed (and predicted) changes are kinetic models. Following the foreword to Benson (1960), here we generalize the notion of kinetic models to any rate process.

2. The process of predicting the changes at operating conditions from the changes empirically observed at accelerated conditions is an extrapolation. Extrapolation inherently carries risk. Thus, to understand the full implications of the extrapolation it is necessary to try to quantify that risk and, where economically feasible, to reduce it.

We call the approach we take to predicting changes in physical and chemical properties physical statistics. The statistics arises from the need to quantify the uncertainties involved in deriving models of (possibly) hidden kinetic processes from empirical data, and then extrapolating with them. The physics arises from the first point listed above. The use of physical statistics was developed to meet the problem of deploying new material systems, with little or no associated operating experience, in applications with extreme reliability requirements. Many engineers engage in versions of this integration of physics and statistics intuitively. In this book, we present the early results of attempting a formal melding of kinetics and statistics, including several new tools for design and analysis of experiments.

1.2 Other Approaches

Before beginning the development of physical statistics concepts and models, it is important to place this particular approach in the context of others used to design and interpret accelerated tests. The physical statistics method, in fact, derives from questions that arise in the application of many of the earlier methods. Below we list five approaches and explain some of the relationships among them.

1. **Classical reliability prediction approach:** According to Pecht and Nash (1994), the classical reliability prediction approach can be traced to a perceived need during World War II to place a standardized figure of merit on system reliability. Component and interconnection reliability were evaluated from accelerated test and field failure data, and combined to obtain a system reliability evaluation. The classical statistical approach described below contributed to the analysis, but the implementation was typically a handbook of failure rates by environment and component/interconnect that was used by designers to determine overall reliability. Inherent in this approach is the assumption that the manufacture of the components was sufficiently standardized to allow the results to be applied across components and that a single, constant failure rate could be used to determine reliability. This later assumption was in part for mathematical convenience and in part a result of the early state of the science.

2. **Classical statistical approach:** This approach arose from the fact that failure times of devices typically appear to have a random component. Statisticians necessarily had to think about these models. The more mathematically inclined statisticians did what all theorists do. They abstracted the problem to make it more solvable. Physical models of acceleration were allowed to take only a few well-defined

forms. In this context, a theory of optimal experiment design and data analysis was developed. Although great strides were made in developing theory around the few models of acceleration that were considered, the class of physical models was unfortunately too restricted. The physical statistical approach arose in part from this approach, with an attempt to consider a much more general class of physical models to develop rules for experiment design and data analysis.

3. **HASS/HALT testing:** This approach has become synonymous with Hobbs (2000). It is sometimes referred to as Environmental Stress Screening/Environmental Stress Testing (ESS/EST). Highly accelerated life testing (HALT) is used to challenge a system during design, and to eliminate by redesign any failures that occur below the fundamental failure limits of the material. Once this design process has been iterated to completion, highly accelerated stress screens (HASS) are put in place to prevent quality problems from affecting the "robustness" of the product. This approach is strongly based on the engineering judgment of the practicing engineer. However, in HALT/HASS, no attempt is made directly to assess the reliability before deployment. One of the strong motivators for some of the theory developed in Chapters 2 and 4 was the challenge of understanding (and developing reliability bounds) for new material systems that have undergone HALT/HASS prior to deployment.

4. **Physics of failure:** This approach, Pecht and Nash (1994), is firmly founded in science, and the foremost proponents of it are the group at Computer Aided Life Cycle Engineering (CALCE) at the University of Maryland. All the known physics about the devices and operating environments are reviewed and modeled to understand where the weak points may appear. Designs and reliability assessments are made using this, and justified with experiment. The approach is complementary to physical statistics. Physics of failure reduces risk from what is known; physical statistics reduces the risk from the unknown.

5. **Physical statistics:** This approach is also firmly founded in science. However, the emphasis, instead of being on known failure modes, is on reducing the risk posed by previously unknown failure modes by using appropriately designed experiments and data analysis that suggests or eliminates from consideration models of the underlying chemistry and physics from accelerated test data.

The purpose of this book is to teach the theory and methods behind physical statistics. The theory and methods described here were developed around problems focused on very high reliability applications, where replacement of parts, when possible, was a multimillion-dollar operation and where new technologies, which had not existed long enough for field-testing, were being employed. The aim was to improve the odds that the deployed systems would last.

Many of the tools of physical statistics are also applicable in less stringent applications. For example, the techniques described in the first part of Chapter 2 can enable the very rapid development of reliability screens for incoming parts. One of the authors developed the screen for incoming parts in 5 min using this theory, and on first application, the screen correctly identified a bad lot of materials, saving the expense of replacing them in the field.

The reader should remember that the focus of physical statistics is reducing the risk of as yet unknown failure mechanisms. The physics of failure approach provides a natural complement in that it provides a focus on failure mechanisms that are known to occur in similar components or systems.

The emphasis on *a priori* unknown failure modes may seem curious to the student. The reason for the emphasis is that, prior to this work, no systematic approach to finding failure mechanisms that attack "under the radar" of standard accelerated test approaches existed. However, periodically such failure modes do occur, and because they often are found after deployment, they can be very expensive to fix. Two published examples of such *a priori* unknown phenomena are as follows:

1. Stress voiding of pure aluminum conductor lines on integrated circuits (Yost et al., 1989). This was discovered by taking integrated circuits, which had been thought reliable, out of storage and finding that they failed immediately, although they had worked when placed into storage.
2. A second, slower degradation mechanism for the degradation of UV-induced gratings in specially doped fiber (LuValle et al., 1998; Chisholm, 1998). This was found by applying methods described in Chapter 2 of this book before deployment. We should note that this is still under investigation and, thus, not definitively known to affect most applications during life.

In both cases, these results were unexpected, and essentially showed that certain material systems assumed to be reliable in a given context were, in fact, not. Because companies tend not to publish their failures, one would expect that in fact many more such phenomena associated with manufacturing modifications that were good for some reason (e.g., cost, efficiency), but that had subtle, bad reliability implications are only known of by the company and the customer.

The primary concern in this book is to develop tools to reduce the risk of these unexpected, *a priori* unknown failure modes. As such, this book complements the philosophy of most other accelerated testing books. In addition, the models discussed in the second half of the book provide a different view of the way models should be built in order to result in reliable extrapolation. These models provide an alternative for data analysis and interpretation.

An interesting side effect of this development is that the theory laid out in these chapters provides a theoretical basis for qualification and screening

Background

tests. These are tests in which devices are subjected to prespecified stress trajectories (sequences of stresses applied over given time sequences) and from which reliability is concluded if no failure or degradation occurs. There are two equally important aspects of screens. Not only must they be effective in detecting defects, but they also must be safely applied so as not to degrade the product seriously. Determining how screens can be safely applied is not easy in practice. One of the demonstrations in Chapter 2 shows, for example, that the test to determine if a thermal cycling screen reduces life is quite different from a rule of thumb that has been used in the industry.

1.3 Foundation of Our Approach

This work is based on two fundamental concepts. Everything that follows can be derived from them.

1. The degradation of performance of any material system designed to do a job (e.g., a laser or coat of paint) is by its nature caused by a change in the properties of that material system over time. By definition, such changes are described by the field of physical chemistry called kinetics. Thus, the appropriate models for such observed degradation should derive from kinetics. For the purpose of this book we define any change in a material, whether it involves making or breaking chemical bonds, transport, or rearrangement of materials, as a kinetic change.

2. The argument of high reliability for a new material system is best made by demonstrating a failure-free period arguably equivalent to operating life. This can be contrasted to the approach of increasing stress to the point that an artificial failure mode is introduced, then proving that that failure mode is artificial. One of the assumed engineering principles for accelerated testing is that higher stress is the best means of inducing and identifying important degradation phenomena. Although it is a necessary first step, in fact, high stress can mask the degradation phenomena important to real-life application, as in the two examples above.

In what follows, it is assumed that the reader has familiarity with the following subjects:

- Physics at the level of *The Feynman Lectures on Physics* (Feynman et al., 1963)
- Chemistry at the level of *The Foundations of Chemical Kinetics* (Benson, 1960) or *Physical Chemistry* (Levine, 1988)

- Statistics at the level of *A First Course in Mathematical Statistics* (Roussas, 1973) or *Mathematical Statistics, a Decision Theoretic Approach* (Ferguson, 1967)

Because most people do not have the opportunity to apply the broad range of tools of these disciplines on a consistent basis, complements to the main chapters provide supplementary material that allows the reader to fill in some of the background.

The following is an artificial example that provides an introduction to several of the concepts we examine later in this book.

1.4 A Simple Example

Consider the case of a new drug in the form of a pill whose shelf life has to be determined. The pill degrades via a reaction with ambient water vapor in the atmosphere, and remains effective as long as >95% of the original formulation is present. Assume that water vapor diffuses through the pill much faster than it reacts; i.e., degradation is limited by the reaction rate. Experiments are run at constant vapor pressure, at 60°C, 80°C, and 100°C. The mean amounts remaining after 1000 h are, respectively, 99.1% 90.0%, and 35.7%. Measurements of the *initial* rates of change show that at 80°C the initial rate is 13 ±0.5 times the rate at 60°C, and at 100°C that rate is 128 ± 4 times that at 60°C (the ± denotes 1 standard error from ten measurements).

The simplest version of reaction rate theory (Glasstone et al., 1941) decomposes a thermally driven reaction rate into two components:

1. A probability or frequency factor, corresponding to the reactants being in the correct configuration for the reaction to occur, and
2. A thermal probability, corresponding to the proportion of reactants having an energy above a certain threshold energy. The threshold probability is determined by Boltzmann's distribution (e.g., Feynman, 1963). Thus, for a given energy threshold $E_{threshold}$, the proportion of molecules with higher energy is $\exp(-E_{threshold}/kT)$, where T is absolute temperature, and k is Boltzmann's constant (for energy measured in electron volts and temperature measured on the Kelvin scale k is 8.625e–5 electron volts (eV)/degree).

If the model is valid, a plot of the log of the initial reaction rate vs. $1/kT$ will give a straight line. The plot is shown in Figure 1.1, where three points representing the data ± 1 standard deviation are shown.

The best straight line (using a least-squares fit) gives a value of ~1.3 electron volts for the "pseudo activation energy." At this point all we know

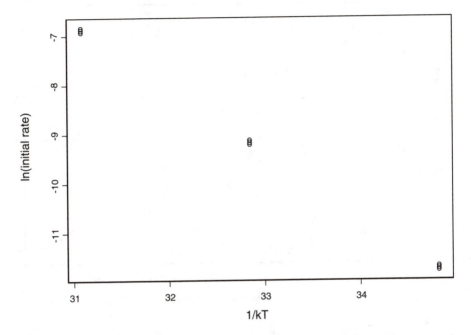

FIGURE 1.1
Arrhenius plot of initial rates at 60°, 80°, and 100°C.

about the reaction is that water is important, and that temperature affects it, but we do not know additional details that can affect degradation as a function of time. We do not know if the reaction occurs as a sequence of reactions involving water first attacking the filler of the medicine, then the principal ingredient (hence "pseudo" activation energy), or if it is a second-order reaction (e.g., Benson, 1960) where two molecules of water must be present simultaneously to cause the degradation. However, we can still obtain some more information from the data we have. If we use the estimated activation energy, we can assume a one-step reaction (although possibly of second or higher order) and calculate an equivalent time for each of the accelerated experiments at room temperature. Assume that vapor pressure of water at each of the different conditions is that for 50% relative humidity at 25°C. Then we can use the estimated activation energy to calculate an acceleration factor at 25°C. The acceleration factor at 60°C, for example, takes the form:

$$A = \frac{\exp\left(\frac{-1.3}{k(273+60)}\right)}{\exp\left(\frac{-1.3}{k(273+25)}\right)} = \exp\left(-1.3\left(\frac{1}{k(273+60)} - \frac{1}{k(273+25)}\right)\right) \approx 203.6$$

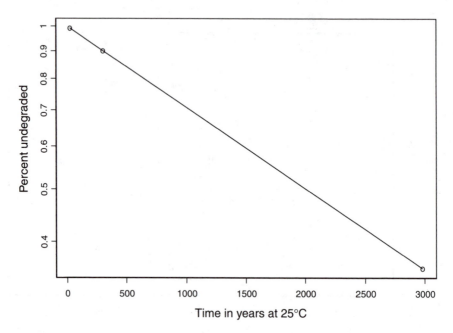

FIGURE 1.2
Fit of observed degradation to an exponential decay, extrapolated through the Arrhenius dependence to 25°C.

Similarly at 80°C the acceleration factor is ~2645, and at 100°C it is 26107.75. The simplest chemical reaction is the first-order reaction, in which there is a finite amount of one material being directly converted to another. In this case, the proportion of starting material remaining (the active ingredient in the pill) decays exponentially. If two molecules react, but the amount of one does not change throughout the reaction (for example, with a steady supply of water vapor to replenish any that has reacted), then the reaction is pseudo first order, and will still decay exponentially. In either case, plotting the log of the proportion remaining vs. time would give a straight line.

This plot at 25°C is given as Figure 1.2. The vertical axis is proportion remaining on a log scale; the horizontal axis is time in years. We see that the assumption of pseudo first order seems to hold and we can read the expected life of the product at 25°C 50% relative humidity. Rough estimation gives about 250 years.

At this point, most reliability engineers would be quite happy with the result. A simple physically plausible model has been developed that fits all of the available data in a definitive manner. If we were doing academic science, this would be where we could end because any additional complications would fall in the realm of further study by other scientists interested in the problem. However, we (the reliability engineering group) are probably the only ones who will investigate this pill before its release. If the effect of underdosing this drug is possibly death, we would be remiss not to conduct a more rigorous study to uncover potentially hidden degradation processes. Thus, there are many more questions to answer, all of which we will attempt to answer at least partly in the remaining chapters.

- How do the uncertainties in the original data translate to uncertainty in the estimates?
- What would we have done with more complicated reactions?
- How do we know that there are no chemical reactions that do not accelerate nearly as much, but that will dominate at operating conditions?

The first question lies in the realm of simple statistical theory, the second in a combination of statistical theory and chemistry, and the third in a combination of economic theory, statistical theory, and chemistry. The solutions involve intelligent combinations of calculation, data analysis, and appropriate experimentation. Unfortunately, because of the relation of specifics of experimentation to the particular area of application (e.g., drug decay, electronics, optics, ecology) this book cannot begin to discuss all the issues necessary to think of in actually performing experiments and making measurements. However, we can discuss mathematical theory associated with the design and analysis of experiments, particularly as they relate to the theory of chemical kinetics, the theory of errors, and the theory of economic trade-offs.

The purpose of this book is to lead the reader through this theory as the authors understand it, via a combination of motivating examples (such as the above) and explicit discussion of the theory. To aid in understanding, code for functions in Splus® that can be used in this analysis is available at the Web site www.crcpress.com/e_products/downloads/, with instructions on its use in the various complements. Readers are encouraged to modify the code as they see fit to improve the approach.

1.5 Organization of This Book

This book is divided into six chapters. This introduction is the first. Chapter 2 introduces the notion of demarcation mapping. Demarcation mapping is a first-order solution to problem 3 above, based on identifying possible simple physical processes that may not have appeared in tests up until now, but may affect life. The demarcation maps are generally based on a simple approximation, and calculations can reasonably done on a hand calculator with a log function. At the end of Chapter 2, the problem of demarcation mapping more complex processes is introduced, and a computational approach to demarcation mapping is developed. Chapter 3 detours to provide a discussion

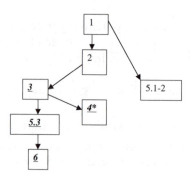

FIGURE 1.3
Chapter dependencies.

of computational issues associated with conveniently specifying and fitting kinetic models, and introduces an approach to computing that the authors have found useful. This sets the stage for most of the methods discussed in the remaining chapters. Chapter 4 provides an extension of demarcation mapping to cover a neighborhood of kinetic processes, around the one that is assumed to correspond to the observations. Such hidden kinetic processes that may appear in operating conditions are called evanescent processes. Thus, Chapters 2 and 4 provide a framework for bounding failure probability, by using accelerated tests in which no observable degradation occurs.

Chapters 5 and 6 change direction. These chapters focus on how to analyze degradation and failure if it occurs. Chapter 5 illustrates how failure time data may be fit; the first part of the chapter introduces an approach that is primarily statistical, but that can be interpreted under special circumstances as kinetic. The modeling is independent of the special software in the first part, but works under more restrictive assumptions. The second part reconnects to the more general framework that arises when one starts with kinetic models. Chapter 6 works through an example fitting a simple set of degradation data. All six chapters have complements for the purpose of providing:

1. Background for the chapter at the level of the references discussed earlier
2. Advanced technical material that does not fit within the main flow of the presentation (e.g., in Chapters 5 and 6, an exercise is provided to lead the reader through some classic accelerated test experiment design questions)
3. Instructions on applying the software available at the Web site to the examples in the book

Figure 1.3 shows the dependence of the chapters on Splus software and on each other. The chapters underlined in the figure require the supplied Splus code, or some equivalent, for implementing the methods described. The early parts of Chapters 2 and 5 can be implemented using Excel® if the

Background

reader so desires. The Splus software is a subset of the code used daily in our work. It is not perfectly transportable to the freeware platform R because of the dependence on certain graphical user interfaces (GUIs), although modification would not be an unreasonable undertaking. We do not provide support for this code, although we do provide free source code. As is normal with shareware, this code is unsupported except for any updates we pull together over time to post at the shareware site.

To use the code, it is necessary to load it into an Splus workspace on the user's computer. The appendix to the book contains instructions on loading the source files directly into Splus by downloading them from the Web site.

Chapter 4 is starred in Figure 1.3 because it contains the newest material. Although the mathematics is more elementary than that supporting Chapters 5 and 6, it is most likely to be unfamiliar; thus, this chapter should be considered an advanced topic.

1.6 Complement: Background Kinetics and Statistics

1.6.1 Arrhenius and Relative Humidity Models

This complement provides an elementary development of the functional dependence on stress of reaction rates for simple chemical reactions. In particular the Arrhenius law of temperature dependence and some laws for humidity dependence are derived.

Derivation of the Arrhenius law: The Arrhenius dependence on temperature is expressed in the formula:

$$\text{rate} = v \exp\left(-\frac{E_a}{kT}\right) \tag{1.1}$$

Here E_a is the "activation energy" for the reaction, k is Boltzmann's constant, T is the absolute temperature, and v is a premultiplier with units of Hertz, representing the frequency with which the reaction may be attempted. In fact, v is also dependent on temperature (e.g., Glasstone et al., 1947), but the dependence is typically so weak compared to the exponential dependence that it can be ignored.

To understand where the exponential function comes from, consider the example of a unimolecular reaction, the simplest type. Consider a molecule in a solution of the chemical methyl cyclohexane. This is a six-carbon ring, joined entirely with single bonds, with a single methyl group (CH_3) attached to one carbon atom. In Figure 1.4 the vertices of the ring are carbon atoms, and each carbon has two hydrogen atoms (not shown) attached. The methyl group is shown explicitly.

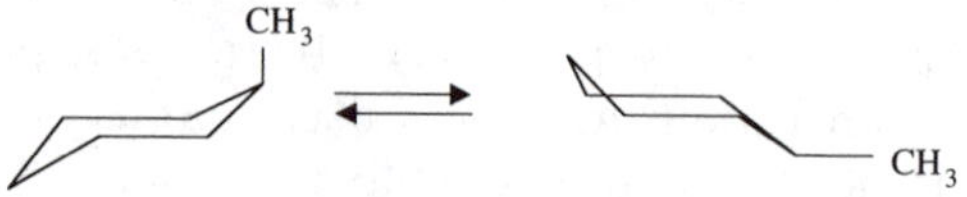

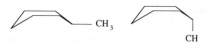

FIGURE 1.4
Axial and equatorial configuration of methyl cyclohexane.

FIGURE 1.5
Intermediate states for transformation of equatorial to axial methyl cyclohexane: "boat" configuration.

These two conformations are interconvertible, but the one on the left (the axial conformation) is slightly less stable than that on the right (the equatorial conformation) due to interference between the methyl group and the hydrogens attached to the carbons in the ring.

From the laws of thermodynamics, the ratio of the two conformations at equilibrium must be that determined by the Boltzmann distribution (e.g., Kittel and Kroemer, 1980, or Feynman, 1963). So we have:

$$\frac{\text{\# of molecules in axial position}}{\text{\# of molecules in equatorial position}}$$

$$= \frac{\exp\left(-\dfrac{\text{energy of axial position}}{kT}\right)}{\exp\left(-\dfrac{\text{energy of equatorial position}}{kT}\right)}$$

$$= \exp\left(-\dfrac{\Delta E(\text{axial - equatorial})}{kT}\right)$$

To transform from either conformation to the other, the molecule must go through an intermediate state of higher energy in which the carbon ring assumes a "boat" configuration as shown in Figure 1.5 (compare to the chair configurations of Figure 1.4).

The proportions in the activated boat states are even smaller than either of the equilibrium states. Assuming no other activated states, the rate of the transformation from axial to equatorial will be of the form:

Background

$$(\text{probability that boat converts to equatorial})$$
$$\times (\text{rate that axial changes to boat}) \exp\left(-\frac{\Delta E(\text{boat} - \text{axial})}{kT}\right)$$

In the reverse direction the rate will be of the form:

$$(\text{probability that boat converts to axial})$$
$$\times (\text{rate that equatorial changes to boat}) \exp\left(-\frac{\Delta E(\text{boat} - \text{equatorial})}{kT}\right)$$

As mentioned above, the premultipliers are temperature dependent as well, as they are functions of the entropy of the system. In summary, the Arrhenius model corresponds physically to the assumption that the dominant temperature sensitive control is by the thermal equilibrium between the starting states and the activated state (boat conformation in the above example).

Relative humidity laws: There are a number of rate laws that have been used for the effect of humidity on reactions (particularly degradation processes). Three that we have found useful are described here. All three are based on the principle that the simplest way for reaction rate to depend on humidity is through the concentration of water at the reaction site. Interestingly, when combined with an Arrhenius law, two of the rate laws are equivalent in the sense that they are statistically indistinguishable when used to fit data. However, their consequences in demarcation maps (Chapter 2) can be quite different. Here, we classify these two together.

Power laws in relative humidity or vapor pressure: An excellent fit of the dependence on temperature of the saturation vapor pressure (SVP) (measured in atmospheres) of water in the atmosphere at sea level is given by the expression:

$$SVP = \exp\left(13.653 - \frac{0.4389}{kT}\right) \quad (1.2)$$

Relative humidity is simply vapor pressure (VP) divided by the saturation vapor pressure, so vapor pressure is

$$VP = RH \times \exp\left(13.653 - \frac{0.4389}{kT}\right) \quad (1.3)$$

A rate model with an Arrhenius dependence on temperature and a power law dependence on vapor pressure can be written:

$$\text{rate} = \alpha V P^\beta \times \exp\left(-\frac{E_a}{kT}\right) = \alpha RH^\beta \times \exp\left(\beta * 13.653 - \frac{\beta * 0.4389 + E_a}{kT}\right) \quad (1.4)$$

This is a power law in relative humidity with an Arrhenius term.

The physical reason for using power laws in vapor pressure is that we assume that the vapor pressure term is proportional to the concentration of water at the reaction site, and the reaction is of order β. In this case a single activated complex requires β water molecules simultaneously to form, which means the rate rises with the β power of the vapor pressure.

Surface and interface interactions: Klinger (1991) following the work of Braunauer et al. (1938) developed an expression for the correct quantity to describe the activity of water, when the rate of reaction is controlled by the formation of monolayers of water at a surface or interface. In particular, expressing relative humidity as a fraction, the appropriate term to use in power laws is the form:

$$\alpha\left(\frac{RH}{1-RH}\right)^\beta \quad (1.5)$$

Like the other models, this can be combined with an Arrhenius model. Note, however, that this will be quite different from the vapor pressure model.

Other models of relative humidity effects exist, each with its niche for application. In each case it is important to keep track of the physics as well as to use careful statistical techniques in fitting the data.

1.6.2 First-Order Kinetics

This complement introduces the notion of linear kinetic models, which apply to reactions designated in chemistry as first-order (or zeroth-order) reactions. A *first-order* chemical kinetic model has the form: $d\vec{A}_t/dt = K\vec{A}_t$, where K is a square matrix and is $\vec{A}_t$ a vector of internal states It is first order by definition because the highest power of any internal state is 1. The simplest first-order system is the system where a substance A is transmuted to a substance B through a chemical change with a single thermal barrier. It can be represented mathematically as the system of differential equations:

$$\begin{aligned}\frac{dA_t}{dt} &= -k_1 A_t \\ \frac{dB_t}{dt} &= k_1 A_t\end{aligned} \quad (1.6)$$

Background

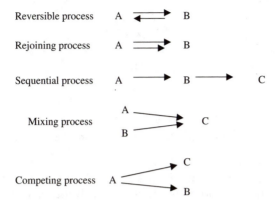

FIGURE 1.6
First-order kinetic models that can be represented with exactly two arrows.

or in matrix form by the equation:

$$\frac{d}{dt}\begin{pmatrix} A_t \\ B_t \end{pmatrix} = \begin{pmatrix} -k_1 & 0 \\ k_1 & 0 \end{pmatrix}\begin{pmatrix} A_t \\ B_t \end{pmatrix} \qquad (1.7)$$

We can also use the following graphical representation for the first-order transmutation of A to B.

$$A \xrightarrow{k_1} B$$

Note that we use a single arrow to represent the reaction because it is a unidirectional transformation with a single thermal barrier. We will designate it as a single arrow process.

The graphical representation and its correspondence to the matrix representation of the first-order chemical kinetic system is very important, because we can interchange the representations when studying the mathematics of the space of kinetic models. For example, using the graphic representation, we can show that there are limits to the ways that a first-order kinetic system can be specified. As an example, it can be shown that there are only five ways to combine single-arrow processes into two-arrow, first-order processes. Each of these ways is shown in Figure 1.6 with a name for simple reference.

The rate constant k_i, for each of these may, for example, be assumed to have the Arrhenius form $k_1 = v \exp(-E_a/kT)$, with v a frequency factory and E_a an activation energy.

To perform general modeling and experiment design, we wish to combine simple models to create more complex models. Programs can be developed to join two matrices, each corresponding to a single-arrow process, to obtain matrices for two-arrow processes. More complicated processes can then be assembled using the same programs. For example, to join two matrices of any size, in the reversible process, we specify equivalence between the beginning state of one matrix with the end state of the other, and vice versa.

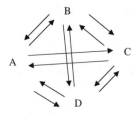

FIGURE 1.7
Fully connected-four state model, simplest model that cannot be generated with the five two-arrow operators.

There are some models that are missed if this approach is used exclusively. An example is the one shown in Figure 1.7 with four states. The above five operators cannot be used to construct this fully connected model.

It is possible to construct this model as long as either B and D or A and C are not connected. To handle this limitation it is necessary to define an operator that connects any two internal states either with a single arrow or reversibly.

A first-order process corresponds to finite starting material flowing though such a process in a conservative manner. A zeroth-order process is one in which at least one of the starting materials is assumed to be essentially infinite, so the effective concentration in that state never changes. A zeroth-order process can be approximated from a first-order process by careful scaling of premultipliers to make the rate of depletion very small.

Integrating a first-order system of differential equations with constant coefficients is actually quite simple. The solution to any system:

$$d\vec{A}_t/dt = K\vec{A}_t,$$

where K is a square matrix and $\vec{A}_t$ is a vector of internal states is just

$$\vec{A}_t = \exp(-Kt)\vec{A}_0,$$

where

$$\exp(-Kt) = \sum_{n=0}^{\infty} (Kt)^n/n!$$

is the matrix exponential. This can be calculated quite efficiently in Splus, so this approach is that used for integration of first-order systems in the freeware program. We assume that changing stress can always be approximated by simple steps in stress. Some source code in Splus is given in complement 1.6.5 for efficient matrix exponentiation.

To connect the kinetic model to data, it is necessary to define a translation program, which turns some of the states into an observable state. For failure time data, the simplest approach is to take a linear combination of states, and assume that failure occurs when that linear combination exceeds some threshold. For example, with a one-step model, the solution to the system of differential equations 1.6 is

$$A_{1t} = A_{10}(1 - \exp(-k_1 t)) \tag{1.8}$$

If we assume that each device has associated a fixed threshold, τ, such that when $A_{2t} > \tau/A_{10}$ failure occurs, and we assume that τ/A_{10} is distributed uniformly and independent of stress, then we have an exponential failure model. The stress dependence is now the stress dependence inherited from the *chemical rate constant* k_1.

A subtle difference in the dimensions of the "exponential" models developed here in the complement and the one found in most statistical texts needs to be emphasized. The latter is an empirical life-stress model developed from a *statistical* analysis of failure (or survival) data. The one developed here is strictly a *kinetic* model describing degradation over time. It is developed assuming first-order kinetics provides a modeling milieu, which is both sufficiently simple and sufficiently close to reality to provide good predictive ability. An exponential distribution for failure arises from assuming a uniform distribution of thresholds to failure in a pure failure model. This is, in principle, testable by microscopic examination. Hence, it is more strictly falsifiable than the empirical exponential model. Not only can it be tested by examining failure time data arising from an accelerated life test, it can in principle be tested by careful physical analysis of failed devices to determine if the material causing failure really does accumulate to a uniformly distributed threshold.

1.6.3 Binomial Distribution and Its Role in Reliability

A very important concept that provides much of the basis of Chapters 2 and 4 is that a reliability experiment that produces no measurable effect (either degradation or failure) can be very informative. Almost as important, but more difficult to handle, are experiments that produce exactly the expected effect. Both kinds of experiment, which provide a null sort of information, are the basis for developing confidence in models used for extrapolation.

A device under test in an experiment will either show a measurable effect or not. A measurable effect could be failure at a given time or a change in some spectrum of properties of the material system in question. However, with no way of knowing *a priori* before an effect is observed what it is, the simplest way to characterize experiments of this type is to use the no effect, some effect dichotomy. If we designate no effect as 0 and some effect as 1 and if we assume that for any given experimental unit there is a probability p, identical across units of a 1 (some effect), then we have described a Bernoulli random variable. If we assume that experimental units are statistically independent, then the sum of the Bernoulli random variables arising from a single experiment with n units will have a binomial (n,p) distribution.

If we denote x as the number of 1s in a binomial sample of n, the probability distribution function for x has the form:

$$P(X = x) = \binom{n}{x} p^x (1-p)^{n-x} = \frac{n!}{x!(n-x)!} p^x (1-p)^{n-x} \qquad (1.9)$$

In other words, Equation 1.9 gives the probability that of the n devices in the experiment, x of them will show a measurable effect. (Note that the meanings of P, p, X, and x are case sensitive.) The uppercase X denotes a random variable, sort of a placeholder for the value of the result of an experiment, before the experiment happens.

1.6.4 Inference for the Binomial Distribution

The experiments of main interest in the Chapters 2 and 4 are those for which the value x of the binomial random variable X is 0 at the end of the experiment, because only then will the experiments indicate support for the model being used for extrapolation. If the binomial random variable is not 0 (e.g., something actually fails or degrades), and the result cannot be identified definitively as experimental error, then a new investigation to understand the possible *evanescent* process detected by the experiment must be conducted. An evanescent process is one not identified in the original set of experiments from which the model used for extrapolation was constructed. In screening during production, the logic is slightly different, but still, a nonzero result requires investigation.

If no devices fail or degrade during the experiment, the simplest estimate of the true value of p is 0. However, there is a finite probability, that for values of $p > 0$, 0 changes occur during the experiment. The question then is how to provide a reasonable upper bound for p. There are essentially two approaches. One is the classical inference method, based on identifying what value of p makes the experiment at least as rare as some very unlikely event. This results in a classical upper confidence bound for p. The other is the Bayesian inference method. In the Bayesian method, the question answered is, if we start with a given prior distribution over the possible values of p, then how should our estimate of the distribution of p change given the experimental results, so any bets we make on p or on the outcome of future experiments remain consistent.

Classical confidence interval for p:

From Equation 1.9 we see that the probability that a binomial experiment with n samples ends up with $X = 0$ is $(1-p)^n$. Thus, the probability that $X \geq 1$ is $1-(1-p)^n$. Note that this latter term increases as p increases. Thus, for a particular n, if we find the value of $\tilde{p}$ such that $1-(1-\tilde{p})^n = .95$, then for every value of p greater than that value, the probability that $X \geq 1$ is greater than 0.95. Thus, we are 95% confident given this experiment that we would see a larger value of X than 0 if $p \geq \tilde{p}$. If we define α so that $(1-\alpha) \times 100\%$ is the confidence level, then the relationship between α, p, and n are defined by the equivalent equations:

$$1-(1-p)^n = 1-\alpha$$

$$(1-p)^n = \alpha$$

$$n = \frac{\ln(\alpha)}{\ln(1-p)} \tag{1.10}$$

$$p = 1 - \alpha^{1/n}$$

When making a business decision, the desire is to choose actions based on expected gains or losses. In effect, we are making a bet. The problem with attempting to use the above formulation to estimate loss can result in logical inconsistencies. These logical inconsistencies can be eliminated by using a Bayesian formulation.

Bayesian inference for P:

Under the Bayesian formulation, we assume that we can specify a prior distribution for p. This is logically equivalent (Ferguson, 1967; DeGroot, 1970) to assuming that we would be willing to place bets on what p is in an internally consistent manner. For the binomial distribution, there is a particular set of distributions that have a very convenient property. They are *conjugate* to the binomial distribution. In particular, if we start with the beta distribution, where we assume p is a random variable, then the beta distribution has the density function:

$$\frac{\Gamma(\alpha+\beta)}{\Gamma(\alpha)\Gamma(\beta)} p^{\alpha-1}(1-p)^{\beta-1}, 0 < p < 1, \; \alpha > 0, \beta > 0 \tag{1.11}$$

The concept behind the Bayesian formulation in this case is that we are willing to bet on different values of p before we see the data, and this allows us to generate a prior distribution. Then an experiment provides further information about p, and if we are to be consistent, we need to modify the prior in a certain way, in particular consistently with Bayes rule. Bayes rule has the following form. Suppose there are events A and B of interest. Then, in its simplest form Bayes rule is

$$\begin{aligned} P(A|B) &= \frac{P(A \cap B)}{P(B)} \\ &= \frac{P(A \cap B)}{P(A \cap B) + P(\neg A \cap B)} \\ &= \frac{P(B|A)P(A)}{P(B|A)P(A) + P(B|\neg A)P(\neg A)} \end{aligned} \tag{1.12}$$

Here $\neg = not$, $\cap$ is the usual intersection symbol, and $P(B|A)$ is the probability of B conditionally given A. In our situation, the probability of $X = x$ given p is just the binomial distribution. Thus, the posterior density of p given x and n is

$$\frac{\dfrac{\Gamma(\alpha+\beta)}{\Gamma(\alpha)\Gamma(\beta)}\binom{n}{x} p^{\alpha+x-1}(1-p)^{\beta+(n-x)-1}}{\displaystyle\int_0^1 \dfrac{\Gamma(\alpha+\beta)}{\Gamma(\alpha)\Gamma(\beta)}\binom{n}{x} p^{\alpha+x-1}(1-p)^{\beta+(n-x)-1}\, dp}$$

$$= \frac{\Gamma(\alpha+\beta+n)}{\Gamma(\alpha+x)\Gamma(\beta+n-x)} p^{\alpha+x-1}(1-p)^{\beta+(n-x)-1} \qquad (1.13)$$

Thus, the posterior density is just a beta with parameters $\alpha+x, \beta+n-x$. The prior can be generalized to a mixture of beta densities, but then both the beta densities and the mixture proportions must be updated.

Assuming that the experiment accurately represents service life for the devices, the worst case we can have from the reliability standpoint is that the 1s represent outright failures. The simplest representation of loss is the loss proportional to failures. So to calculate the expected loss given the result of an experiment, we just need calculate the expected probability of failure. In the beta distribution this is simply

$$\int_0^1 p \frac{\Gamma(\alpha+\beta)}{\Gamma(\alpha)\Gamma(\beta)} p^{\alpha-1}(1-p)^{\beta-1}\, dp = \frac{\alpha}{\alpha+\beta} \qquad (1.14)$$

This is the expected value of p based on the assumed (prior) distribution for p. The experiment result gives us the posterior (conditional) distribution to calculate the expected value of p. From Equation 1.14, the posterior expected loss given no failures during an experiment with n units is $\alpha/(\alpha+\beta+n)$. Thus, suppose we start with a beta prior distribution with parameters α and β before an experiment, and after an experiment with n units and no failures occur. Then we would accept either side of a bet with payoff α if no failure occurred, and payoff $\beta+n$ if a failure occurs in an $n + $ first trial of the same experiment. In other words, to be consistent with the data, we have adjusted the odds of failure from $\alpha{:}\beta$ before the experiment to $\alpha{:}(\beta+n)$ after the experiment.

1.6.5 Splus Source Code for Matrix Exponentiation

Source code for efficient matrix exponentiation in Splus is provided here. The term in quotes with an arrow pointing at it is the name of an Splus object, while the string to the next term in quotes with an arrow next to it is the object. Thus "matexp.pck" is the string vector containing the strings "matexp.pck," "matexp1," "matpow1," "bindec0," "bindeci0." These strings are the names of the functions that are used in the matrix exponentiation program. The function "matexp1" is the master matrix exponentiation function. This code is optimized for conservative first-order kinetic processes (e.g., solutions that are stochastic).

```
"matexp.pck "<-
c("matexp.pck," "matexp1," "matpow1," "bindec0," "bindeci0")
"matexp1"<-
function(M)
{
#
#This function calculates exp(M) where M is a
#square matrix
#
#First normalize the matrix so that calculating the
#exponential does not cause computational difficulties
#
        M1 <- diag(length(M[, 1]))
        val <- max(abs(M))
        val1 <- (floor(val)) * length(M[1, ])
        val1 <- max(val1, 1)
        Ma <- M/val1
        M2 <- M1
        n <- 1#
#Calculate the value of exp(M(normalized))=eM
#
        repeat {
                M2 <- M2%*% (Ma/n)
                n <- n + 1
                M1 <- M1 + M2
                if((sum(abs(M2)) < 10^-9))
                        break
        }
#
#Raise eM to the power used to normalize it originally
#
        M1 <- matpow1(M1, val1)
        M1
}
"matpow1"<-
function(M, val)
{
#
#This program calculates the value of the square matrix M raised
#to the integer power val.
#It is assumed that M is stochastic
#Determine the binary decomposition of val
#
        xvec <- bindec0(val)
        mb <- M
```

```
              vec <- c(xvec[length(xvec)], diff(rev(xvec)))
              n <- length(vec)
              m2 <- diag(length(mb[, 1]))
              m1 <- mb
              if(n >= 1) {
                     for(i in 1:n) {
                            x1 <- (vec[i])
                            if(x1 >= 1) {
#
#Recursively square the matrix as many times
#as necessary
#
                                   for(j in 1:x1) {
                                     m1 <- m1%*% m1
                                   }
                            }
#
#Multiply the squared products
#
#correct m1 to be stochastic
                                   n1 <- length(m1[1, ])
                                   for(j in 1:n1) {
                                     m1[, j] <- m1[, j]/sum(m1[, j])
                                   }
                                   m2 <- m2%*% m1
                     }
              }
              m2
}
"bindec0"<-
function(n)
{
#
#this program calculates the binary decomposition of a given number n
#by using the iterative program bindeci0
#
       v <- NULL
       ind <- T
       nx <- n
       while(ind) {
              dum <- bindeci0(nx, v)
              v <- c(v, dum$xout)
              ind <- dum$xind
              nx <- dum$xnext
       }
       v
}
"bindeci0"<-
function(n, v)
{
#
#this program calculates the largest integer power of two that
#fits into a number, n, and stashes that calculation into the
#vector v. If the remainder is greater than 1 it hands the remainder
#to itself recursively and continues. Note that the log(n,2)
```

```
#functionality is only available through Splus 2000, in splus 6.0 and
#beyond it is necessary to use log(n)/log(2) in its place to get the
#logarithm base 2 of a number
#
      na <- log(n)/log(2)
      nb <- floor(na)
      n1 <- (n - (2^nb))
      if(n1 < 1) {
            out <- nb
            nextx <- 0
            ind <- F
      }
      else if(nb != 0) {
            out <- nb
            nextx <- n1
            ind <- T
}
else {
      out <- 1
      ind <- F
      nextx <- 0
      }
      list(xout = out, xnext = nextx, xind = ind)
}
```

2

Demarcation Mapping: Initial Design of Accelerated Tests

Accelerated testing is a common tool in reliability studies. In its classical implementation, it works especially well with devices of intermediate reliability such as InGaAs laser diodes. With accelerated testing we can uncover the (known) failure modes and obtain precise estimates of reliability, so we know how much redundancy to build in to make a reliable system. However, for ultrahigh-reliability devices, the use of accelerated testing becomes more problematic. We attempt to deduce the reliability of a device (i.e., lack of failure or degradation) over a very long period of time at operating stress, from a short exposure to high stress. Any degradation or failure that is observed during the accelerated test that is relevant during operating life disqualifies the device. Failure or degradation that represents an artificial failure mode provides no useful information about the reliability of the device under the conditions in question and, in fact, could mask a relevant failure mode.

If failure and degradation do not provide information on reliability (except in cases where an enormous amount is already known about the failure physics of the device under test), then where is the information on reliability in accelerated tests? *The information on reliability in an accelerated test of a very high reliability device is in the stress-time regimes free from and prior to any degradation or failure.* This concept is rather radical for most practitioners and theorists involved in accelerated tests, but it is inescapable if one allows for the possibility of one degradation or failure mode masking another.

For most new devices, there is some prior experience with similar devices, hence some expected failure or degradation modes. There is also, however, always the possibility of something new. So truly rigorous testing may require both characterizing the acceleration of known failure modes (or screening them out if possible), and reducing the risk from unknown failure modes. Characterizing the acceleration of known failure modes is the standard problem in accelerated testing, so the theory in texts such as Nelson (1990) or Meeker and Escobar (1998), possibly augmented by models and methods such as described in Chapters 3, 5, and 6 of this book, is sufficient.

However, both for the development of screens that do not reduce the remaining life significantly, and for reducing risk of *a priori* unknown failure modes, it is necessary to have in place a theory from which it is possible to infer reliability from the initial failure and degradation-free periods in accelerated testing.

This chapter provides an introduction to some theory and methods necessary for inferring reliability from failure and degradation-free periods. The statistical theory underlying this development was begun in Complements 1.6.3 and 1.6.4. The physical theory and the integration of the two occupy most of this chapter and Chapter 4. In this chapter we introduce the notion of demarcation maps, both analytic (LuValle et al., 1998; LuValle, 2000) and computational (LuValle et al., 2002). In conjunction with this development, we show how demarcation maps can be used to accomplish the following:

1. Design and interpret acceptance tests: Acceptance tests are accelerated tests where no failures during the test are meant to imply something positive about reliability. An example of an acceptance test can be given for the case of a device with a solder joint that has a low (e.g., 100°C) melting temperature. The problem is to specify the combination of time and temperature that will be at least as hard on the device as the design lifetime exposure under service conditions, with the constraint that the acceptance test temperature must be below 100°C if it is to reflect design lifetime conditions.
2. Design and interpret safety tests for screens: Safety tests are tests that examine the residual life of devices that survive a screen meant to eliminate early failures.
3. Design and interpret certain step stress experiments: Here the step stress experiments are designed to probe for failure and degradation modes that are masked by failure or degradation modes observed at high stresses.

Demarcation map theory allows study of one specified mechanism with unknown parameters. Evanescent process maps, covered in Chapter 4, allows this to be generalized to the study of neighborhoods of mechanisms. In this chapter and in Chapter 4, both the power and ultimate limitations of accelerated testing are exposed. One result of full understanding of both of these chapters is that there is always some risk involved from the reliability point of view in introducing a new material system to a new operating environment. Understanding these two chapters gives the reliability practitioner the ability to articulate that risk explicitly in terms of probability and physical mechanism, and to systematically reduce that risk.

Demarcation maps can be divided into two categories: *analytical* and *computational*. These are described below:

1. *Analytical* demarcation maps: These are developed by constructing an analytical approximation to determine which among the set of potential chemical reactions with given parametric form are complete. Given

the approximation and the stress conditions, we can divide the set into two regions. One region is where all the potential reactions are essentially completed (the region of potential failure modes) and the other where they are not. By comparing these maps for two different stress trajectories, an accelerated stress trajectory, and an operating trajectory, four regions are identified: (a) a region where the potential reactions are not complete for either stress condition, (b) another where they are complete for accelerated test conditions but not for operating life conditions (the region of potential artificially generated failure modes), (c) another where they are complete for operating life but not for accelerated test conditions (a region of potential failure modes not uncovered by the accelerated test), and (d) a region where they are complete for both stress conditions.

2. *Computational* demarcation maps: The concept is the same as that of the analytical, but rather than make analytical approximations to determine when reactions are complete, the extent of potential reactions is calculated for each stress trajectory and compared directly at each point. The dominating trajectory is indicated at that point by symbol or color choices.

The latter is, in fact, more flexible and lends itself to both more complex kinetic processes, and more complex statistical analysis. However, it also requires more sophisticated software for doing the calculations. The former is more easily calculated, and offers surprisingly useful insights. We will begin this chapter studying the tools and theory of analytical demarcation maps.

Although at present it seems impossible to make a detailed map of this sort for all chemical reactions, it is possible to construct a fairly conservative map for a moderate-sized set of chemical reactions (LuValle et al., 1998, 2000). The theory supporting the conservative nature is fairly involved, but the basic idea is straightforward and is provided in the next section. Following that are examples for several kinds of degradation processes, listed below. The first three are based on the assumption of simple, one-step processes or empirical models. The fourth example uses computational demarcation mapping with more complicated kinetic processes.

For simplicity, in the first seven sections of this chapter, where the physical theory of demarcation mapping is developed, we ignore the question of sample size and statistical interpretation. We return to statistical theory in Sections 2.8 and 2.9.

2.1 Analytical Theory of Thermal Demarcation Maps

The demarcation approximation was originally developed in the context of material systems with a macroscopically observable response driven by a set of parallel thermally activated (Arrhenius) first-order kinetic processes (see Complement 1.6.1). It is assumed that each process consists of a single step and there is a distribution of activation energies across processes (Primak, 1955). With these conditions the rate constant for each local process has the form:

$$k_1 = v \exp\left(\frac{-E_a}{kT}\right) \qquad (2.1)$$

where v is a premultiplier, k is Boltzmann's constant, T is absolute temperature, and E_a is the activation energy (in the following, all activation energies are given in electron volts, eV). The relative amount (normalized to total amount that can react, also called the reaction *extent*) reacted over time for this local process then is (see Complement 1.6.2 for the form of the solution to first-order kinetic reactions):

$$(1 - \exp(-k_1 t)) \qquad (2.2)$$

An interesting thing happens if, for a given time and temperature, we plot the value of Equation 2.2 given Equation 2.1 across activation energies, and compare that to the approximation:

$$\Delta(k_1, t) = \begin{cases} 0, k_1 t < 1 \\ 1, k_1 t \geq 1 \end{cases} \qquad (2.3)$$

Figure 2.1 shows the result (actually plotting $1 - \exp(-k_1 t)$ and $\Delta(k_1, t)$ for a given temperature and time against activation energy). The center vertical line is the approximation. The circles are the exponential functions at each activation energy. The two outer vertical lines correspond to values of v two orders of magnitude off. The bounds shown by the outer lines are important in the following discussion. Relationships 2.1 and 2.3 give the formal approximation:

$$E_d = kT \ln(vt) \qquad (2.4)$$

where E_d is the bounding value for activation energy, or the demarcation energy. Reactions with activation energies less than E_d are essentially complete for the given conditions of T, t, and v.

Equation 2.4 is the thermal demarcation approximation. The fundamental point of interest is that this approximation describes a relationship between all values of E_a and all values of v for given values of T and t. Essentially Equation 2.4 describes an approximate front between values of the pair (E_a, v) corresponding to reactions that are finished, and values of the pair corresponding

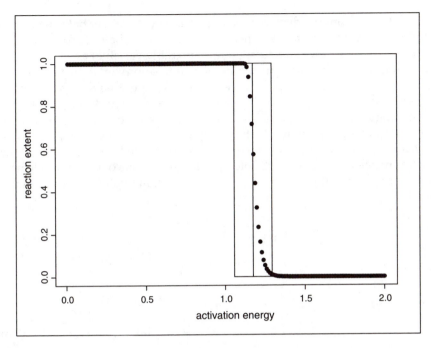

FIGURE 2.1
Extent of reaction vs. activation energy plotted over the demarcation approximation.

to reactions that have not begun. In a plot of log(v) (vertical) as a function of E_a (horizontal), reactions represented by paired values of E and v in the region above and left of the line are complete and those below and right are not. The term demarcation mapping derives from this combination of the demarcation "line" (or more generally "curve" as shown below) on the plot ("map") of v vs. E_a. Note that this description is now independent of the need to consider only physical processes with distributions of activation energies.

In the original use of the demarcation approximation as a method of scaling time (Primak, 1955; Erdogan et al., 1994), the uncertainty was associated with the distribution of activation energies within a material. Its validity was driven by the law of large numbers, acting through the number of sites in the material participating in macroscopic effects. The use of the demarcation approximation promoted here (LuValle et al., 1998, 2000; LuValle, 2000) the uncertainty now lies in the human lack of knowledge of the mechanism. Certain bounds may apparently be used to ensure a conservative assessment (LuValle, 2000) although these need either to be checked on a case-by-case basis or to be understood from a general mathematical theory. Certainly, the treatment of parameters here leans toward a Bayesian approach. At the end of this chapter we follow this idea further, developing a generalized theory for acceptance testing based on looking for processes shadowed by observed degradation processes.

With this notion of a demarcation approximation, we see that it would be of great interest to the test engineer to be able to see how the demarcation approximation for the time–temperature (possibly variable) exposure corresponding to life compares to the demarcation approximation for a given accelerated test. A good *a priori* accelerated test would be one that reached the same set of (E_a, v) pairs during the test, as are reached during life. Before examining maps of such pairs over given stress exposures, we need to know how to deal with Equation 2.4 when temperature is changed at a given time, as happens both in life and in step stress accelerated testing. To see how such a change is calculated, we note that at a given value of v we can calculate times and temperatures corresponding to the same value of E_a by simply writing:

$$kT_1 \ln(vt_1) = E_d = kT_2 \ln(vt_2) \tag{2.5}$$

or

$$t_1 = \frac{(vt_2)^{T_2/T_1}}{v} \tag{2.6}$$

which implies that, after a step in temperature, Equation 2.4 can be re-expressed:

$$E_d = kT_2 \ln\left(v\left(t_2 + \frac{(vt_1)^{T_1/T_2}}{v}\right)\right) \tag{2.7}$$

A key practical issue is how low the value of v can be. As far as we have found, there is no scientific reason to presume a lower bound. Diffusion processes using this approximation can be expected to reach very low values. However, for the one-step processes, quoted values typically range from 10^6 to 10^{18} Hz. Based on our experience with glasses and polymers, we have chosen to use 10^{-5} Hz as a lower bound, but some careful thought needs to be used. Even with this uncertainty, the maps provide significant information on the relative efficacy of different accelerated tests. A reasonable, conservative alternative for thermal demarcation maps is to extend the maps to low enough values of v so that the corresponding E_d is 0 in all experiments and life. In the example below this would be to 10^{-10}.

To illustrate how demarcation mapping works, consider an operating life consisting of a 4-h bake at 180°C, followed by 25 years at 50°C. We compare this to an accelerated test with a 4-h bake at 180°C followed by 1 h at 260°C in Figure 2.2. (Figure 2.2 and Figure 2.3 were produced using the code in the complement to this chapter in Splus.) The shaded region (horizontal

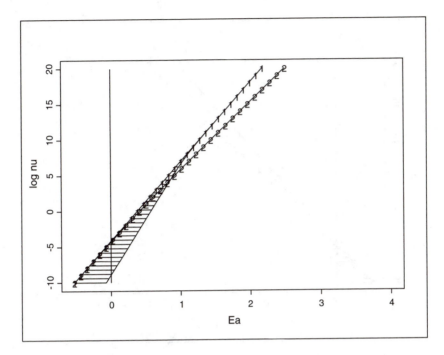

FIGURE 2.2
Demarcation map of 180°C 4 h + 50°C 25 year life vs. 180°C 4 h + 260°C 1 h experiment.

lines) represents the region of reactions corresponding to life (it does not include the burn-in as that occurs before the customer sees the device, and failures in that part of the space will be eliminated by the burn-in). The space between the lines labeled with 1s and that labeled with 2s represents the region of reactions corresponding to the accelerated test. We note that the accelerated test contains all reactions occurring during life only for values of ν greater than $10^{4.5}$ Hz. The vertical line is at 0 eV, representing the limit of "reasonable" reactions. If we have reason to regard any reaction with ν less than this to be an improbable contributor to failure, we can regard survival of the accelerated test without failure as a basis for acceptance of the product. If the value of ν might actually be close to 10^4 Hz, we may wish to use the bounds on the approximation given by the outer lines in Figure 2.1 to guide us. Thus, we might wish to use an experiment such that the accelerated test region contains all reactions occurring during life for values of ν greater than 10^2 Hz. Figure 2.3 shows an experiment doing this. However, instead of having to extend the time on test by a factor of 100, we only had to test at 260°C for 10 h, rather than 1 h.

A schema for an Excel spreadsheet is supplied as Table 2.1. The first six rows are adjustable parameters; the remaining rows are formulae for columns.

We can also note for these diagrams that the accelerated tests include many reactions not included during life. Thus, we see with this particular scheme that there is a fair danger of seeing an artificial failure or degradation mode

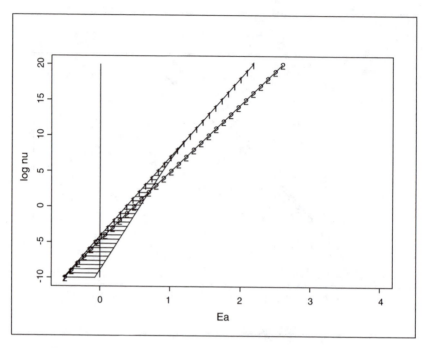

FIGURE 2.3
Demarcation map of 180°C 4 h + 50°C 25 year life vs. 180°C 4 h + 260°C 10 h experiment.

during accelerated testing that would have nothing to do with service life failure. The demarcation maps can be used to design tests with more specificity as well, trading off time on test, with coverage, and the possibility of a false alarm by using different time-stress trajectories. The maps can even be used to design tests that peel back degradation or failure modes that analysis indicates are artificial so we can look for failure mechanisms they may be masking. A prior publication (LuValle, 1998) provides an example of this in a real problem. Here we continue this example to show in principle how this could work.

Suppose that exposures of 4 h at 180°C and 10 h at 260°C result in several failures, and that repeating the experiment with the second exposure at 300°C also results in failures. Suppose further that we have recorded the times of these failures and find that they correspond to a process with an activation energy of ~2.0 eV and a v value of 10^{15} Hz. Because a failure mode with these parameters will not occur during service life, the standard assumption applied in accelerated testing is that we now have proved the device reliable, because the only failure we have seen is one that will never occur during life. However, the map points out that our accelerated tests have not gotten close to a number of (E_a, v) pairs that may be active during life. The high-stress failure mode could be masking it in these tests. Thus, what we want to do is to perform a test that minimally activates the (2.0, 10^{15}) failure mode, but that emphasizes some of the lower values we have missed. Such a test is described below.

TABLE 2.1

Adjustable Terms	Variable Name	Example Value
Operating temperature	OT	85
Operating time	Otime (h)	24*365*25
Experimental temperature	ET	260
Experimental time	Etime (h)	1
Bake temperature	BT	180
Bake time	Btime (h)	4

Column Name/Symbol	Column Formula	Example Value
$\mathrm{Log}_{10}(v)$	lv	5
v	10^{lv}	100000
Demarcation energy (bake)	$(8.625e-5)(273+BT) \times \ln(v \times \mathrm{Btime} \times 3600)$	0.8239
Equivalent time under operating conditions, t_{bO}	$\dfrac{\left(v \times \mathrm{Btime} \times 3600\right)^{\frac{(273+BT)}{(273+OT)}}}{v \times 3600}$	1077.3
Equivalent time under experimental conditions, t_{bE}	$\dfrac{\left(v \times \mathrm{Btime} \times 3600\right)^{\frac{(273+BT)}{(273+ET)}}}{v \times 3600}$	0.1688
Demarcation energy (operating life)	$(8.625e-5)(273+OT) \times \ln\left(v \times \left(\mathrm{Otime} + t_{bO}\right) \times 3600\right)$	0.9882
Demarcation energy (experiment)	$(8.625e-5)(273+ET) \times \ln\left(v \times \left(\mathrm{Etime} + t_{eO}\right) \times 3600\right)$	0.9129

The experiment below is possible if the device under question is very small, so that its thermal inertia is very low (it can heat up and cool down very quickly). This is possible with some microelectronic components, and some optical fiber components. In Figure 2.4, life stress is as before. The experimental stress is 180°C for 4 h, followed by 320°C for 15 s, followed by 170°C for 500 h. The shaded region again represents the one-step, first-order reactions that can occur during life. The 320°C exposure is the region between the line labeled 1 and the line labeled 2. It pulls the final exposure out close to the failure mode regime (the F) but just short of it. The remaining exposure, 170°C for 500 h, provides significant aging of the device for most failure mechanisms corresponding to the shaded regime, for values of v down to $\sim 10^{-6}$ Hertz. (where the activation energy is >0).

This exposure is illustrative of the potential of step-down stress tests to explore regimes that might typically be masked in highly accelerated tests. However, it is also illustrative of some of the cautions. In particular, this sort

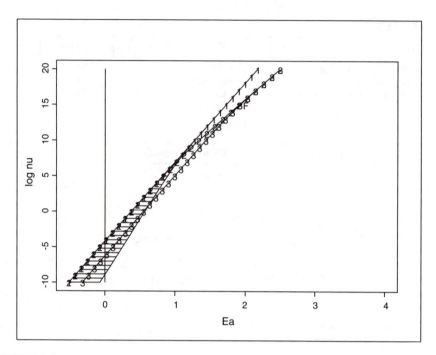

FIGURE 2.4
Demarcation map of 180°C 4 h + 50°C 25 year life vs. 180°C 4 h + 320°C 15 s + 170°C 500 h experiment.

of stress test would be impossible for a device with high thermal mass and low surface area to volume. The map also points out that some potential failure modes are economically untestable (very low ν values).

This section has illustrated some of the basic concepts of the demarcation map, using the demarcation approximation derived from the Arrhenius relationship. In the next section, we look more closely at the particular problem of developing an acceptance test for Arrhenius-driven failure modes.

2.2 Designing an Acceptance Test for a Purely Thermal Process

If our goal is to design an acceptance test (a test such that if nothing fails we will accept that the device is reliable), we would like the acceptance test to provide coverage for all the processes we believe might occur during life, while minimizing the chances of rejecting a good material system, with the minimum cost and time expenditure. The plots and discussion above suggest a single number that we can extract from the demarcation maps, which provides a simple characterization of the degree of conservativeness of the accelerated test with respect to accelerable failure modes. That number is

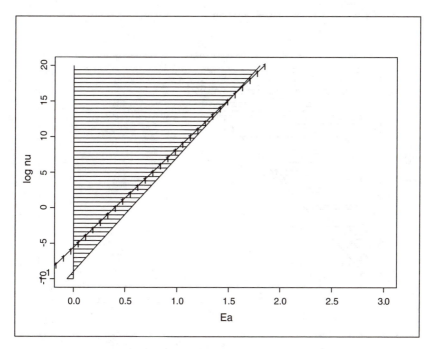

FIGURE 2.5
Demarcation map of 40°C 30 year life vs. 90°C 100 h experiment.

the minimum value of ν where the accelerated test contains all the processes occurring during life. This is the simplest use of demarcation mapping to design accelerated tests that will arguably reach end of life. A conservative value of ν for the purpose of including both processes running totally to completion and some more complex processes (LuValle, 2000) is a value three orders of magnitude lower than the lowest value for a simple single-step chemical process in the materials under consideration.

Suppose the device to be life-tested is assembled using a low-melting-point solder (e.g., indium–tin with a melting point of 110°C). We believe that the device, which will see a maximum temperature of 40°C during manufacturing and life, may have a thermally activated failure model. If we would like to check whether the device could survive for 30 years, while testing at no more than 90°C, how long will we have to test?

Constructing a demarcation map using the theory provided above, we can construct a demarcation map for 40°C for 30 years (no bake) and plot over it the demarcation map for 90°C 100 h.

The shaded region in Figure 2.5 contains the (E_a, ν) pairs that can be expected to react during life, for those values of ν in the range of 10^{-10} to 10^{20}. The region to the left and above the line marked with "1"s is the corresponding region for the experiment. We see that the value of ν for which the activation energies activated during the experiment are exactly those activated during life is 10^{16}. This is fairly high, as many common chemical

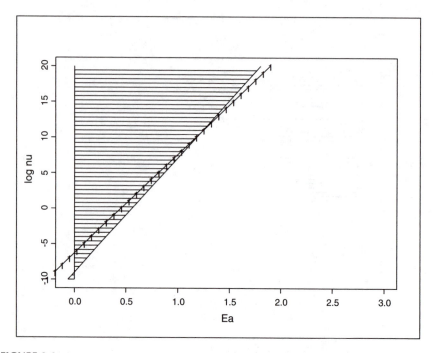

FIGURE 2.6
Demarcation map of 40°C 30 year life vs. 90°C 500 h experiment.

reactions have ν values below 10^{16} Hz. A good moderately conservative value would be 10^3 or lower.

Plainly, to have a more conservative test we need to lengthen the time at 90°C. If we change the time at 90°C to 500 h we obtain Figure 2.6, with a crossover of approximately $10^{10.5}$. The reader can continue with trial and error. Alternatively, the reader can complete the following exercises.

Exercise 2.2.1

Determine how to identify for fixed operating time and temperature and experimental temperature the exact time needed to obtain equivalent times at a given value of ν.

Exercise 2.2.2

Prove that for fixed temperature experiments, with the experiment at a higher temperature than the operating temperature, this experiment will be conservative for all values of ν larger than that found in 1.

Exercise 2.2.3

An alternative approach used in reliability engineering is to find the smallest activation energy such that, for every activation energy above that energy, the experiment is an end-of-life experiment. Prove that the demarcation energy corresponding to the crossover value of v is exactly this activation energy.

2.3 Simple Temperature/Humidity Models

In the simplest form, the rate of reaction for combined temperature/humidity models might be assumed to have the form:

$$k_1 = vf(RH, T)\exp\left(\frac{-E_a}{kT}\right) \quad (2.8)$$

where

$$f(RH, T) = \begin{cases} (RH)^\beta & \text{[PL]} \\ (Vapor\_Pressure)^\beta & \text{[VP]} \\ \left(\frac{RH}{1-RH}\right)^\beta & \text{[BHT]} \end{cases} \quad (2.9)$$

depending on the physics. The bottom expression was derived to model the number of monolayers of gas adsorbed on a surface (Braunauer et al., 1938; Klinger, 1991), and has been found to be useful in modeling reactions rates occurring at material interfaces. The other two are used often, although careful derivations of which situations are most appropriate for which model are not available. As remarked in Complement 1.6.2, the first and second representations are not statistically identifiable when coupled with the Arrhenius relationship. Equation 2.8 results in a definition of demarcation energy with the form:

$$E_d = kT\ln\left(vf(RH,T)t\right) \quad (2.10)$$

From this we can derive a similar result to that given in Equations 2.5, 2.6, and 2.7, so that after a step in stress, the demarcation energy may be calculated:

$$E_d = kT_2 \ln\left(v\left(t_2 + \frac{\left(vf(RH_1, T_1)t_1\right)^{T_1/T_2}}{vf(RH_2, T_2)}\right)\right) \quad (2.11)$$

2.4 Designing an Acceptance Test for a Temperature/Humidity Model

Suppose that we start with a device that we suspect has a degradation mode accelerated by both temperature and humidity (or that we want to argue does not have any relevant failure mode so accelerated). Further suppose that each device is given an 85°C, 85% relative humidity (RH), 100 h screen by the customer, and then the device goes into a well-controlled telecommunications central office (maximum long-term temperature 45°C, maximum long-term humidity 42%.) for 20 years. How long and under what conditions do we have to test to reach approximate end of life assuming a minimal v value of 10^3?

The freeware program provides facilities to construct an array of demarcation diagrams looking at each of these functions. For example, testing an experiment run for 500 h at 90°C, 90% RH, with values of β 1 or 2 (reaction rate proportional to concentration of water, or proportional to square of concentration) results in the plots in Figure 2.7.

The lines without "+" show operating life (recall there is a burn-in defining the beginning of operating life, hence two lines), while the lines with "+" are experiments. The analytical solutions again are conservative although finding each one becomes difficult. The partial vapor pressure in these plots was calculated using the empirical relationship (Equation 1.3):

$$VP = RH * \exp\left(13.653 - \frac{0.4389}{kT}\right).$$

For the power law it is clear that we have not reached the goal proposed above. The reader can input the software in the complement, and proceed by trial and error until a solution is found, or the reader can perform the following exercises:

Exercise 2.4.1

Determine how to identify for fixed humidity function operating time and temperature, relative humidity, β, and experimental temperature and relative humidity the exact time needed to obtain equivalent times at a given value of v.

Exercise 2.4.2

Prove that for fixed stress experiments, with the experiment at a higher temperature than the operating temperature, this experiment will be conservative for all larger values of v found in 1. What happens for the humidity functions?

Demarcation Mapping: Initial Design of Accelerated Tests

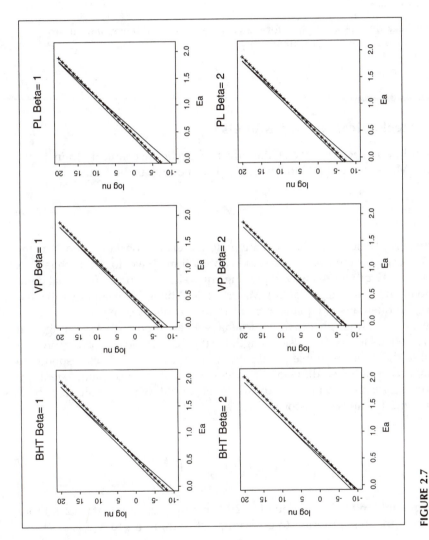

FIGURE 2.7
Array of temperature/humidity demarcation maps of life consisting of 85°C 85% RH screen followed by 20 years at 40°C 42% RH vs. an experiment of 90°C 90% RH for 500 h.

Exercise 2.4.3

What advantages do the plots have over the crossover points in terms of information about artificial and uncovered failure mechanisms in accelerated tests?

Exercise 2.4.4

Check the crossover point for the vapor pressure and power law curves. Which is more conservative? Why?

2.5 Mechanical Cycling Models

A simple empirical model of the effect of cyclic mechanical strain is the Coffin–Manson model (Manson, 1981). The model has the form:

$$\varepsilon_p = M N_f^z \qquad (2.12)$$

where ε_p is plastic strain, M and z are material-related constants, and N_f is the number of cycles the device withstands until failure. If we do not reach failure, Equation 2.12 provides a way of judging equivalent amounts of internal damage to the material (degradation). Miner's rule (Miner, 1945) supports this use of the Coffin–Manson equation. Empirically for notched metal samples, z is close to $-\tfrac{1}{2}$, although other values have been reported. Assuming that ε_p corresponds to a self-diffusion length for the material on each cycle (proportional to the nonrecoverable plastic strain), then z close to $-\tfrac{1}{2}$ corresponds to failure occurring when diffusion has occurred to a certain constant extent.

Ignoring for the moment the diffusion argument, which would fix z, then Equation 2.12 can be transformed to

$$1 = \frac{\left(M^{1/z} N_f\right)}{\left(\varepsilon_p\right)^{1/z}} \Leftrightarrow \left(\frac{1}{z}\right) = \left(\frac{1}{\log(\varepsilon_p)}\right) \log\left(M^{1/z} N_f\right) \qquad (2.13)$$

The equation on the left of Equation 2.13 corresponds to Equation 2.4. From this the step stress criterion to go from plastic strain ε_{p1} to ε_{p2} is

$$N_{f2}^* = \frac{\left(M^{1/z} N_f\right)^{\log(\varepsilon_{p2})/\log(\varepsilon_{p1})}}{M^{1/z}} \qquad (2.14)$$

Miner's rule (Miner, 1945) gives empirical support to the use of Equations 2.13 and 2.14 to construct demarcation maps with varying amounts of plastic strain. Our implementation of a demarcation map plots $\log_{10}\left(M^{1/z}\right)$ vs. $\left(-1/z\right)$. In this way, the plot can be interpreted the same way as the thermal demarcation map, with values above and to the right of the drawn line values that will occur during the stress trajectory.

The maps are drawn in the $\left[\log_{10}\left(M^{1/z}\right), \left(-1/z\right)\right]$ plane. However, the value of $z = -1/2$ is marked with a red line to aid in interpreting the plot in line with the diffusion hypothesis. Understanding other values of z would require additional theoretical analysis.

2.6 Acceptance Testing for Mechanical Cycling Induced by Thermal Cycling

Typically, in electronic equipment, mechanical cycling is done indirectly, either through thermal cycling, vibration, or power cycling of the devices. Thus, to employ this calculation directly some side experiments must be done to ascertain the relationship between the applied cycle and the plastic strain seen in the device under question. Further, in chemistry, values of v are sometimes estimated experimentally, and there is some theory (Glasstone et al., 1941) relevant to the task. It is not clear that values of M or $M^{1/z}$ are similarly attainable anywhere. The main assumption in what follows is that the primary effect of the thermal cycle is that the mismatch between the coefficients of thermal expansion for different materials drive a mechanical cycling in a material seeing significant plastic strain. The chemical effects due to raised temperature are assumed negligible.

Suppose that we have a problem where a device might be expected to be turned on once every hour for 2 years in normal operation, and the plastic strain applied to the component of interest is 1e–4 each time it is turned on. To ensure that very bad devices are not shipped, the device might undergo ten thermal cycles corresponding to a plastic strain of 1e–3. Following a standard rule-of-thumb engineering practice for developing such environmental stress screens that we have encountered, suppose that 20 devices were subjected to 100 thermal cycles, and none failed.

The question a conservative reliability engineer would ask is, were the 100 thermal cycles enough to make sure we have not increased the failure rate by applying this test? As a first cut we can specify values of $M^{1/z}$ to be used in the plot of $10^{-5} - 10^{10}$.

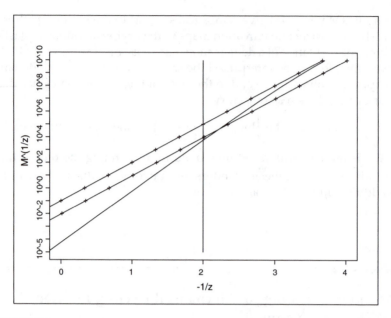

FIGURE 2.8
Coffin–Manson demarcation maps of life consisting of 10-cycle screen at 1e–3 plastic strain followed by 2 years of hourly cycles at 1e–4 plastic strain vs. an experiment of a 10-cycle screen at 1e–3 plastic strain followed by 100-cycle "safety test" at 1e–3 plastic strain.

The map, given in Figure 2.8, shows that even if we consider only the value of $(-1/z)$ corresponding to diffusion, the "safety experiment" of 100 thermal cycles has not tested to end of life, so does not indicate if we will see any serious increase in failure due to this testing or not. The line without "+" is the line representing the screen followed by life, while the lower line with "+" is the screen followed by the safety test; the top line is the screen. The vertical line is the value of $(-1/z) = 2$.

As two alternatives, we can consider thermal cycles with identical strain as the screen, and thermal cycles at a lower strain level than the screen. Figure 2.9 runs the test to 170 cycles. Now the safety test at 1e–3 plastic strain does reach end of life under the diffusion hypothesis, but not for more conservative values of –1/z. Figure 2.10 shows what happens if we drop the plastic strain in the safety test to 3e–4 for 500, 2500, and 5000 cycles (respectively, the second, third, and fourth lines with "+" from the top).

Exercise 2.6.1

Determine algebraically how many cycles are required in the safety experiment to ensure no failures by end of life assuming a – 1/z value of 2, and of 1, for a plastic strain value of 3e–4.

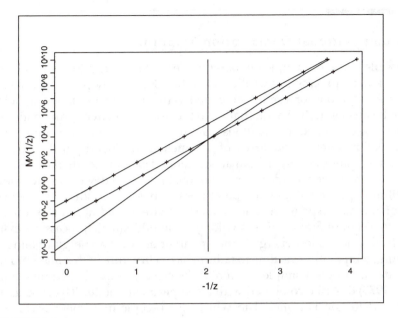

FIGURE 2.9
Coffin–Manson demarcation maps of life consisting of 10-cycle screen at 1e–3 plastic strain followed by 2 years of hourly cycles at 1e–4 plastic strain vs. an experiment of a 10-cycle screen at 1e–3 plastic strain followed by 170-cycle "safety test" at 1e–3 plastic strain.

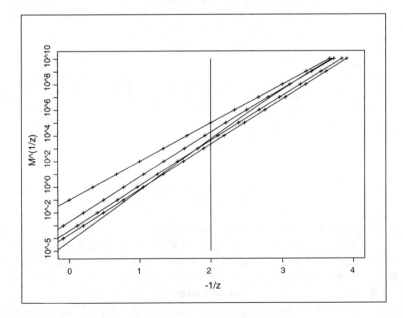

FIGURE 2.10
Coffin–Manson demarcation maps of life consisting of 10-cycle screen at 1e–3 plastic strain followed by 2 years of hourly cycles at 1e–4 plastic strain vs. an experiment of a 10-cycle screen at 1e–3 plastic strain followed by 3e–4 plastic strain at 500, 2500, and 5000 cycles.

2.7 Computational Demarcation Mapping

It is possible to use the simple temperature demarcation map to bound more complex kinetic processes (LuValle, 2000), but that is only proved for constant stress situations for a small set of models (the two-arrow kinetic models that can be accelerated). An alternative to the use of elegant approximations such as those contained in Equations 2.4 through 2.7 is to perform a numerical calculation, over a large array of parameters and identify if the extent of reaction of interest is larger from an accelerated experiment than from life.

For an example we consider a problem that appears in devices made from metal films deposited at high temperatures on substrates with lower coefficients of thermal expansion. The problem is stress voiding, and is often related to electromigration. Stress voiding is an odd kinetic process, in that the "effective activation energy" (the apparent activation energy obtained by plotting temperature vs. time to failure) has a curvature with a maximum. The model we use is a simple version of the creep process that Krauss and Erying (1975) describe combined with a compartment model. The model of Krauss and Erying is a model in which microscopic movement involves overcoming an activation barrier, and stress preferentially changes the activation energy. In particular the model has the form:

$$A_1 \underset{k_b}{\overset{k_f}{\rightleftarrows}} A_2 \underset{k_b}{\overset{k_f}{\rightleftarrows}} \ldots \underset{k_b}{\overset{k_f}{\rightleftarrows}} A_n \quad (2.15)$$

where

$$k_f = (\text{fluence})^{\text{pow}} v \exp\left(-\frac{E_a - \text{stress.coef} \times (T_0 - T)_+}{kT}\right) \quad (2.16)$$

and

$$k_b = (\text{fluence})^{\text{pow}} v \exp\left(-\frac{E_a + \text{stress.coef} \times (T_0 - T)_+}{kT}\right) \quad (2.17)$$

T_0 is the temperature where the film and the underlying material are at 0 stress, the notation $(z)_+$ denotes we are setting the value as z if it is positive, and 0 if it is negative. We could take an arbitrarily complex kinetic model if we wished, but this is sufficiently complex. This model could be used as a combined model of electromigration and stress voiding, with fluence denoting current density. We assume there is no fluence effect. Then in the simplest version of the real problem, there are four unknown parameters to cause us concern: $(v, E_a, \text{stress.coef}, T_0)$, and there is a moderately complex relationship based on Equations 2.15 through 2.17.

Demarcation Mapping: Initial Design of Accelerated Tests

Strictly, this sort of model is a local microscopic model of one side of a region where the void is forming. Hence depletion of the early states corresponds to formation of the void. Thus, a good surrogate for the actual degradation is the concentration of material left in the first few states.

To evaluate this model, we need to proceed in four steps:

1. Write a program to evaluate the degradation model, including both kinetic equations, and its translation to an observable at given parameter values for given stress trajectories. Our approach uses a finite set of compartments.
2. Write a program to step through the parameters, evaluating a stress trajectory, and recording the observable.
3. Compare observables for life and experiment trajectories.
4. Iterate until a combination of experiments is decided.

Steps 1, 2, and 3, can be done in environments such as Mathcad®, Mathematica®, and Matlab®. For demonstration purposes, the kinetic modeling software described in the complement to Chapter 3, written in Splus is used.

The model we use in studying this problem is of the form from Equations 2.15 through 2.17 with 20 states. Total depletion of the first five states is taken as the kinetic marker for degradation or failure.

We assume for this device that we are concerned only with the time the device is actually on the shelf, and that when it is in use it runs hot enough so that there is 0 stress. We assume that the shelf life is 5 years. The parameter values we actually do the mapping for are as follows:

$$\log_{10}(v) = (-4, 0, 4, 8, 12, 16, 20)$$

$$E_a = (0.3, 0.6, 0.9, 1.2, 1.5)$$

$$T_0 = (200, 350, 500), \text{ or } 600$$

$$\text{stress.coef} = (1e-6, 1e-5, 1e-4, 1e-3, 1e-2)$$

T_0 varies through 200, 350, and 500 for the values in life. In six experiments it follows this variation, and in the remaining experiments it is set to 600. This allows us to examine the effect of T_0 both if it is uncontrolled (but constant through manufacture), and if it can be used as an experimental variable for acceleration.

Figure 2.11 is a demarcation map comparing storage life at 50°C for 5 years to an accelerated experiment at 150°C for 5000 h. Each graph in the plot corresponds to the values of pow (the exponent of the fluence term = 0 identically), T_0, and stress.coef being fixed to the values shown above that graph for the life stress. In this case the T_0 values changed in the experiment in a corresponding fashion. The horizontal axis of each graph is the activation

TABLE 2.2

Evaluation of Potential Experiments

Temperature	Time	T_0	Number (of 525 possible)
100	5000 h	200,350,500	372
150	5000 h	200,350,500	363
100	2 years	200,350,500	357
150	2 years	200,350,500	350
200	2 years	200,350,500	405
300	2 years	200,350,500	413
50	2 years	600	196
100	2 years	600	194
150	2 years	600	195
200	2 years	600	269
100*	1000 h	600	218
250*	1000 h	600	270
Combo (two above*)			190

energy (E_a) in electron volts, and the vertical axis is $\log_{10}(v)$. An L in a spot means that life will result in more depletion in the first five compartments than the experiment, so the experiment cannot be thought of as having reached end of life. An E implies that the experiment will result in more depletion, implying that for the parameter values corresponding to that position we have reached beyond the end of life degradation in the experiment. (A key assumption here is that end of life is controlled only by the function of the state vector we are examining, and the environment influences end of life only through that state vector).

A very simple crude measure of the efficiency of an experiment is the number of points in the map where life dominates experiment. For a fixed map, such as here, the fewer points dominated by life, the more likely the experiment is to identify regions where failure can occur. Although this actually must be weighed by the likelihood of each parameter point, for the interest of simplicity we start with the simple measure and allow Chapter 4 to consider more sophisticated measures. Table 2.2 shows the experiments tried vs. the number of points where life dominates.

There are several points of interest that can be derived from this table. First, if T_0 can be controlled, it makes a far more effective accelerant than temperature for probing for this phenomenon. Second, increasing T_0 shifts the optimal temperature for achieving high coverage. Finally, combining experiments cleverly can result in significant shortening of the time required to attain a certain level of coverage.

The particular coverage by the 2-year experiment at 100°C with a T_0 of 600°C is shown in Figure 2.12 and can be contrasted to the coverage shown in Figure 2.13 for the combined experiment run at 1000 h each. The differences are actually slight, although they do exist.

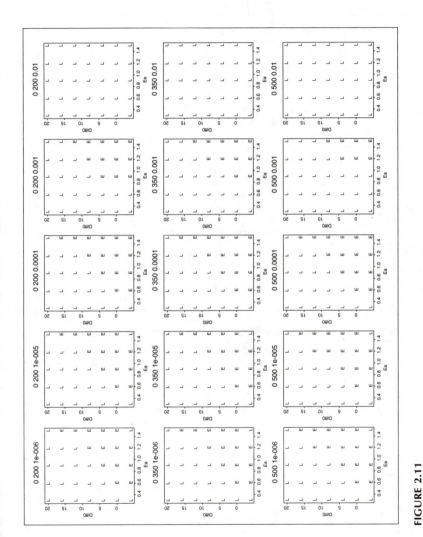

FIGURE 2.11
Computational demarcation map of stress voiding, comparing life at 50°C for 5 years vs. an experiment at 150°C for 5000 h. L means that the reaction extent for life dominates the reaction extent for the experiment at that combination of parameters. E means the reverse.

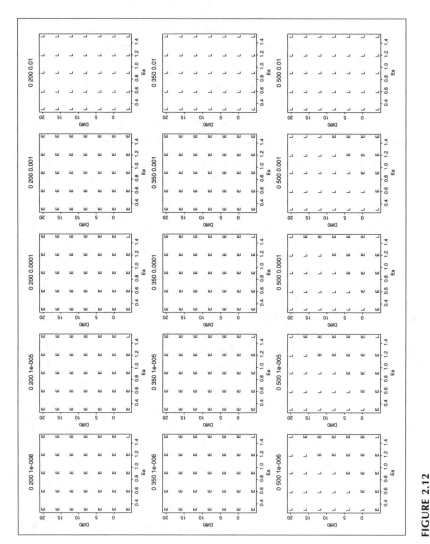

FIGURE 2.12
Computational demarcation map of stress voiding, comparing life at 50°C for 5 years vs. an experiment at 100°C for 2 years, with $T_0 = 600°C$.

Demarcation Mapping: Initial Design of Accelerated Tests 49

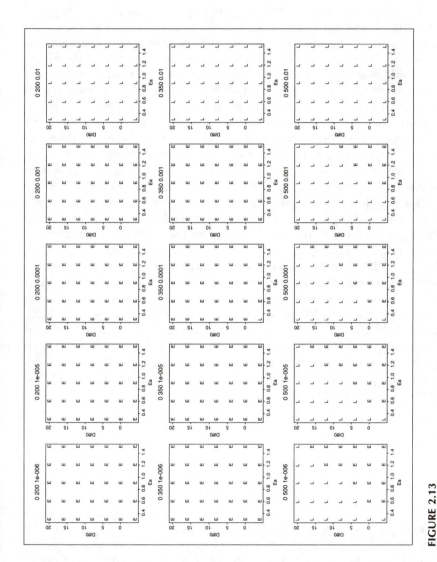

FIGURE 2.13
Computational demarcation map of stress voiding, comparing life at 50°C for 5 years vs. combined experiment 100°C for 1000 h, and another experiment of 250°C for 1000 h, both with $T_0 = 600$°C.

2.8 Beta Binomial Interpretation of 0 Failures

Consider Figure 2.12. This figure represents an experiment where some number of devices were aged for 2 years at 100°C. For each point with an E in the diagram, if the devices obey an aging law corresponding to that point, then the experiment aged each of those devices beyond the level where they would be aged during life. Thus, for no failures during the experiment, at those points where an E sits in the diagram, we can assume that the experiment provides an upper bound for failure during life. From Complement 1.6.4, if we start with a $\beta(\alpha,\beta)$ prior distribution for each point, then at each point with an E, our posterior distribution would be a $\beta(\alpha,\beta+n)$. Thus, at each point with an E, the odds we would consider a fair bet for failure have changed from $\alpha:\beta$ (corresponding to an $\alpha/\alpha+\beta$ probability of failure) to $\alpha:\beta+n$ (corresponding to an $\alpha/\alpha+\beta+n$ probability of failure). On the other hand, at each point where an L occurs, there has been less aging than would occur during life. Assuming that Murphy's law rules the universe, then the monotone function relating failure and degradation to the state vector could be sharply concave upward, so the noticeable degradation or failures could be waiting until toward the end. Thus, where an L is written, we have no information on bounding the probability of failure during life.

Ultimately, the purposes of our accelerated tests are to inform the people in the business/operation area how to make a decision. Thus, these diagrams must be converted somehow to a number that gives us the ability to decide between a yes and a no. A good number to base that on is the probability of failure during life. To reach that in a consistent way, we have to extend the Bayesian paradigm (Ferguson, 1967; DeGroot, 1970) that we have been using. We have to assign a prior probability to each point in the hyper-rectangle in Figure 2.11. Then before the experiment, assuming no failures, the probability of failure is

$$\int_H \int_0^1 P \frac{\Gamma(\alpha+\beta)}{\Gamma(\alpha)\Gamma(\beta)} P^{\alpha-1}(1-P)^{\beta-1} dP d\pi(\phi) = \frac{\alpha}{\alpha+\beta} \int_H d\pi(\phi) = \frac{\alpha}{\alpha+\beta} \quad (2.18)$$

The term $\pi(\phi)$ is simply the probability distribution over the hyper-rectangle H depicted in Figure 2.11; ϕ is the point representing a particular set of the parameters. Letting E denote the region of H where the experiment dominates life and letting L denote the region where life dominates the experiment, by our argument above, the posterior probability that a failure or noticeable degradation will occur during life after an experiment has occurred is

Demarcation Mapping: Initial Design of Accelerated Tests

$$\iint_{E\ 0}^{\ 1} P \frac{\Gamma(\alpha+\beta)}{\Gamma(\alpha)\Gamma(\beta+n)} P^{\alpha-1}(1-P)^{\beta+n-1} dP d\pi(\theta)$$

$$+ \iint_{L\ 0}^{\ 1} P \frac{\Gamma(\alpha+\beta)}{\Gamma(\alpha)\Gamma(\beta)} P^{\alpha-1}(1-P)^{\beta-1} dP d\pi(\theta) \quad (2.19)$$

$$= \frac{\alpha}{\alpha+\beta+n} \int_E d\pi(\theta) + \frac{\alpha}{\alpha+\beta} \int_L d\pi(\theta)$$

$$= \frac{\alpha}{\alpha+\beta+n} \times \pi(E) + \frac{\alpha}{\alpha+\beta} \times (1-\pi(E))$$

From the final portion of Equation 2.19 we see that the best experiment (disregarding costs) is the one with the highest prior probability on the points where the experiment will dominate life. Thus, the best experiment to reduce the risk associated with assuming no failures will occur during life is the experiment most likely to disprove the hypothesis that no failures will occur during life. This result is important enough to emphasize it.

2.9 An Extrapolation Theorem

Assume the observable, y, is Bernoulli with probability $p(\phi, E_i)$. Assume the following:

1. That $p(\phi, E_i)$ is unknown, but that there is a known ordering induced by the experiments. Thus, if $E_i \neq E_j$, then knowing ϕ implies that we know if $p(\phi, E_i) \leq p(\phi, E_j)$, $p(\phi, E_i) = p(\phi, E_j)$, or $p(\phi, E_i) \geq p(\phi, E_j)$. This assumption is the key assumption tying the statistical theory developed here to the substantive science. In the situation described in this chapter, the theory of computational demarcation maps is what provides us with the ability to make this statement.
2. That there are desirable and undesirable outcomes (e.g., survival and failure), and the undesirable outcomes correspond to $y = 1$.
3. That for each experiment E_i there is a corresponding sample size n_i of the Bernoulli y.
4. That $L_\theta(\phi) = n_p p(\phi, E_p)$. Here $L_\theta(\phi)$ is the loss if ϕ is true, and we extrapolate using θ. E_p is the special experiment corresponding to what we are trying to extrapolate to. Define the risk of an experiment $R(E_i)$ as the expected loss given the prior on the ϕ.

5. That costs for all E_i (except E_p) are constant.
6. That $p(\theta, E_i) \equiv 0$ for all i.

Assume that at each point ϕ a beta prior is specified for $p(\phi, E_i)$ and assume that it is the same beta prior at all ϕ.

Define $S_{ij} = \{\phi | p(\phi, E_i) \geq p(\phi, E_j)\}$, and Θ_θ as the set of all the ϕ in the neighborhood of θ that we wish to consider. Define an estimate determined from E_i for $p(\phi, E_j)$ to be a strongly determined upper estimate only if $\phi \in S_{ij}$. Then we have:

LEMMA
Under the assumptions, and assuming $n_i \equiv n$, the experiment that, with a null result (no observable degradation or failure), minimizes risk using strongly determined upper estimates is that experiment for which $\pi(S_{ip})$ is maximized.

PROOF

$$R(E_i) = E(L_\theta(\phi) | E_i) = n_p \int_{\Theta_\theta} \int_0^1 dp\, p(\phi, E_p) \beta(\alpha, \beta | E_i, \phi) d\pi(\phi) \quad (2.20)$$

Here the second equality follows from assumption 4, and $\beta(\alpha, \beta | E_i, \phi)$ is the beta posterior density given experiment E_i and ϕ. Using the strong determination criteria, this is a $\beta(\alpha, \beta + n)$ density over S_{ip}, and a $\beta(\alpha, \beta)$ density over $\Theta_\theta \setminus S_{ip}$. Thus, under strong determination we can decompose Expression (2.20) into

$$R(E_i) = n_p \left(\left(\frac{\alpha}{\alpha + \beta + n} \right) \int_{S_{ip}} d\pi(\phi) + \left(\frac{\alpha}{\alpha + \beta} \right) \int_{\Theta_\theta \setminus S_{ip}} d\pi(\phi) \right)$$

$$= n_p \left(\left(\frac{\alpha}{\alpha + \beta + n} \right) \pi(S_{ip}) + \left(\frac{\alpha}{\alpha + \beta} \right) (1 - \pi(S_{ip})) \right) \quad (2.21)$$

Clearly, Expression 2.21 is minimized by maximizing $\pi(S_{ip})$. Note this is exactly the described result. In particular, S_{ip} is precisely the set of points where no signature events during experiment i strongly implies an upper bound on the probability of a signature event during E_p.

2.10 Summary

In this chapter a radically different approach to the extrapolation of accelerated tests has been described. This theory is based on the following notion: *The information on reliability (for ultrahigh-reliability devices) in an accelerated test is in the failure and degradation free periods prior to any observable degradation or failure.* We have provided a theoretical basis for using this approach assuming that tests can be constructed in which no failures are expected to occur given the existing models of failure, and assuming that the failure modes are described by a rather simple one-step kinds of kinetic models.

The complement to this chapter provides instructions for using the downloadable computer code in Splus for analytical demarcation mapping.

In Chapter 4, we extend the notion of demarcation mapping to considering more complex (but still first-order) chemical kinetic processes as possible alternatives to whatever model currently fits the failure and degradation data obtained.

2.11 Complements to Chapter 2

The purpose of this section is to provide examples of using the software for analytical demarcation maps. Examples of use of the software for computational demarcation maps must wait until the complement for Chapter 3, when the kinetic modeling software is provided. Complement 2.11.1 contains an extension of the extrapolation theorem in Sections 2.8 and 2.9 to multiple experiments.

2.11.1 Demarcation Maps for Multiple Experiments

In most real problems we will wish to conduct an array of experiments. The lemma in Section 2.9 only deals with choosing a single experiment. The calculation of posterior probabilities for multiple experiments is relatively straightforward. In particular, the strong determination criterion defines a partition on the space of parameters into $S_{i\rho}$ and $\Theta_\theta \setminus S_{i\rho}$ for each experiment i. For n experiments this gives us partitions:

$$\begin{pmatrix} S_{1\rho} = S_{1\rho}^1 & \Theta_\theta \setminus S_{1\rho} = S_{1\rho}^0 \\ S_{1\rho}^1 & S_{2\rho}^0 \\ \cdots & \cdots \\ S_{1\rho}^1 & S_{m\rho}^0 \end{pmatrix} \quad (2.22)$$

A complete partition of Θ_θ into mutually non-intersecting sets can be constructed from this. In particular define:

$$P_k = \bigcap_i S_{ip}^{binary(k,i)}$$

where binary(k,i) is the ith digit in the binary representation of k, k from 1 to 2^m. The posterior probability for the m NULL experiments, the ith experiment with sample size n_i, from which risk can be calculated is

expected posterior probability of failure

$$= \left| \sum_k \frac{\alpha \times \pi(P_k)}{\alpha + \beta + \sum_{i=1}^{m}(binary(k,i) \times n_i)} \right| \qquad (2.23)$$

With an array of fixed environmental settings, the problem of optimization becomes a problem of distributing samples between the different experiments. For simplicity assume a fixed total number N of experimental devices, then Equation 2.23 can be re expressed:

expected posterior probability of failure

$$= \left| \sum_k \frac{\alpha \times \pi(P_k)}{\alpha + \beta + \sum_{i=1}^{m}(binary(k,i) \times Np_i)} \right| \qquad (2.24)$$

where p_i is a proportion, and

$$1 = \sum_{i=1}^{m} p_i$$

Ignoring for the moment that devices come in whole numbers, we can apply elementary calculus. Define from the constraint

$$p_m = 1 - \sum_{i=1}^{m-1} p_i$$

Then the partial derivatives take the form:

$$\frac{\partial}{\partial p_i}(\text{expected posterior probability of failure}) =$$

$$N \left(\left(\sum_k \frac{\text{binary}(m,i) \times \alpha \times \pi(P_k)}{\left(\alpha + \beta + \sum_{i=1}^{m}(\text{binary}(k,i) \times Np_i)\right)^2} \right) - \left(\sum_k \frac{\text{binary}(k,i) \times \alpha \times \pi(P_k)}{\left(\alpha + \beta + \sum_{i=1}^{m}(\text{binary}(k,i) \times Np_i)\right)^2} \right) \right)$$

We leave it as an exercise to the interested reader to show that the common 0 point for this derivative is a minimum given the constraints on the p_i.

2.11.2 Using the Freeware

The purpose of this section is to acquaint the user with the code associated with this section. We make primitive use of the native GUIs available in Splus in this section, to provide simple drop-down menus and popup commands. The Appendix to the book describes how to install the freeware.

Once all the code is available in Splus, the simplest way to use it is using the drop-down menu and the primitive GUIs native to Splus. For those who have used SPLIDA™ coded by Bill Meeker, the interface here will seem primitive. We encourage ambitious readers to modify the source code to their convenience. However, it would be wise to retain the original as a backup.

If the reader has followed the instructions in the appendix, there should be an entry in the top toolbar "Demarcation maps." Left-clicking the mouse button on this entry gives a drop-down menu with the list

Life-emphasized temperature demarcation map
Temperature/humidity demarcation map
Coffin Manson demarcation map
Create function for direct kinetic evaluation using kinetic model
Plot extent matrices from direct kinetic eval
Plot INT of extent matrices from direct kinetic eval

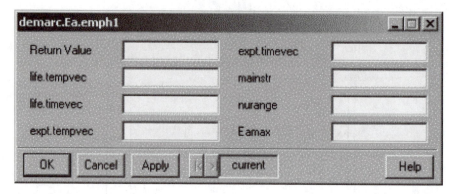

FIGURE 2.14
GUI for thermal demarcation map function in Splus.

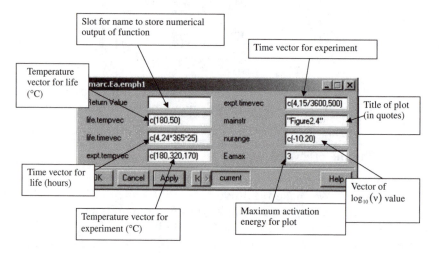

FIGURE 2.15
Annotated GUI for thermal demarcation map function in Splus, filled in to produce Figure 2.4.

The last three commands are connected to evaluating computational demarcation maps, so we will wait until the complement to Chapter 3 to see how to use them. The first three items are the three analytical demarcation maps. The simplest way to check them is against the calculations presented in the book.

2.11.2.1 Thermal Demarcation Maps

To access the GUI for simple analytical thermal demarcation maps, left-click on "Demarcation maps" above the toolbar in Splus, and then left-click on the "Life emphasized temperature demarcation map" entry in the demarcation map drop-down menu. This brings up the default GUI for the function "demarc.Ea.emph1," shown in Figure 2.14. In Figure 2.15 we show this GUI filled out to construct Figure 2.4 in the text, and provide a guide to each slot.

Demarcation Mapping: Initial Design of Accelerated Tests 57

As configured, left-clicking with the mouse on the apply button causes the function to execute, producing Figure 2.4 and bringing up the "report" window in Splus, where the numerical output is dumped. The numerical output is a list, printed below. Note that vectors in the input use the usual Splus notation beginning with a c(...).

```
$order.1.life:
      matx   mat.ret    vec.ret
 [1,]   -5 -0.07571782 0.2499654
 [2,]   -4  0.01424706 0.3141129
 [3,]   -3  0.10421194 0.3782609
 [4,]   -2  0.19417682 0.4424103
 [5,]   -1  0.28414170 0.5065632
 [6,]    0  0.37410657 0.5707249
 [7,]    1  0.46407145 0.6349086
 [8,]    2  0.55403633 0.6991479
 [9,]    3  0.64400121 0.7635266
[10,]    4  0.73396608 0.8282502
[11,]    5  0.82393096 0.8938027
[12,]    6  0.91389584 0.9612132
[13,]    7  1.00386072 1.0322211
[14,]    8  1.09382560 1.1086037
[15,]    9  1.18379047 1.1906019
[16,]   10  1.27375535 1.2766537
[17,]   11  1.36372023 1.3649038
[18,]   12  1.45368511 1.4541597
[19,]   13  1.54364999 1.5438388
[20,]   14  1.63361486 1.6336898
[21,]   15  1.72357974 1.7236094
[22,]   16  1.81354462 1.8135564
[23,]   17  1.90350950 1.9035141
[24,]   18  1.99347437 1.9934762
[25,]   19  2.08343925 2.0834400
[26,]   20  2.17340413 2.1734044

$order.1.expt:
      matx   mat.ret     vec.ret    vec.ret
 [1,]   -5 -0.07571782 -0.07568411 0.1107292
 [2,]   -4  0.01424706  0.01430510 0.1987238
 [3,]   -3  0.10421194  0.10431185 0.2867194
 [4,]   -2  0.19417682  0.19434876 0.3747162
 [5,]   -1  0.28414170  0.28443745 0.4627144
 [6,]    0  0.37410657  0.37461487 0.5507146
 [7,]    1  0.46407145  0.46494374 0.6387176
 [8,]    2  0.55403633  0.55552947 0.7267251
 [9,]    3  0.64400121  0.64654624 0.8147398
[10,]    4  0.73396608  0.73827349 0.9027669
[11,]    5  0.82393096  0.83113808 0.9908161
[12,]    6  0.91389584  0.92574114 1.0789056
[13,]    7  1.00386072  1.02282347 1.1670707
[14,]    8  1.09382560  1.12311511 1.2553813
[15,]    9  1.18379047  1.22708066 1.3439780
[16,]   10  1.27375535  1.33470647 1.4331434
[17,]   11  1.36372023  1.44550735 1.5234262
```

```
[18,]    12   1.45368511    1.55874689  1.6158006
[19,]    13   1.54364999    1.67368908  1.7116992
[20,]    14   1.63361486    1.78974188  1.8125674
[21,]    15   1.72357974    1.90648713  1.9189315
[22,]    16   1.81354462    2.02365199  2.0299552
[23,]    17   1.90350950    2.14106672  2.1441137
[24,]    18   1.99347437    2.25862872  2.2600638
[25,]    19   2.08343925    2.37627697  2.3769438
[26,]    20   2.17340413    2.49397556  2.4942834

$crossover:
[1] 0.8400869  0.4496849
```

The component order.1.life is the matrix of demarcation points corresponding to life. The first column is $\log_{10}(v)$, the second is the demarcation energy corresponding to that value of v after 4 h at 180°C. The third is the demarcation energy after a subsequent 25 years at 50°C.

The component order.1.expt is the similar matrix for the experiment temperature trajectory. The component labeled crossover is the estimated values of E_a and $\log_{10}(v)$, where the curve representing the end of the experiment and the curve representing the end of life cross over one another.

2.11.2.2 Temperature/Humidity Demarcation Maps

To access the GUI for simple analytical temperature/humidity demarcation maps, left-click on the "Temperature/humidity demarcation map" entry in the demarcation map drop-down menu. This brings up the default GUI for the function "array.rhplot," shown in Figure 2.16 filled in to produce Figure 2.7.

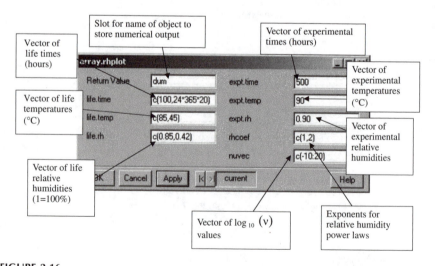

FIGURE 2.16
Annotated GUI for temperature/humidity demarcation map function in Splus, filled in to produce Figure 2.7.

Left-clicking the apply button with the mouse produces the plot below, as well as assigning the numerical output to the name "dum." The output is printed below.

The numerical output is a list; at the highest level it has two components, corresponding to each exponent. Within each list is a sublist with components BHT (the Teller model RH/(1-RH)), VP (the vapor pressure model), and PL (the power law in relative humidity model). Even though there is no numerical difference between the last two if a translation is made for activation energy and v, both are shown because there is a possibility that one is interpretable when the other is not because, for example, one results in a negative activation energy. Each of the sublists has a cc component and a matrix component. CC shows the crossover value of $\log_{10}(v)$; while the matrix is similar to that for temperature, only the life matrix and experiment matrix are shown side by side. The column of $\log_{10}(v)$ values is repeated at the beginning of both matrices, life first.

```
> dum
[[1]]:
[[1]]$BHT:
[[1]]$BHT$cc:
[1] 3.250845

[[1]]$BHT$mat:
         matx    mat.ret   vec.ret     matx vecx
 [1,]     2    0.5907987  0.6745286    2 0.6639237
 [2,]     3    0.6618967  0.7381145    3 0.7360148
 [3,]     4    0.7329948  0.8018350    4 0.8081058
 [4,]     5    0.8040929  0.8657280    5 0.8801969
 [5,]     6    0.8751910  0.9298391    6 0.9522880
 [6,]     7    0.9462890  0.9942218    7 1.0243790
 [7,]     8    1.0173871  1.0589357    8 1.0964701
 [8,]     9    1.0884852  1.1240441    9 1.1685611
 [9,]    10    1.1595832  1.1896091   10 1.2406522
[10,]    11    1.2306813  1.2556852   11 1.3127433
[11,]    12    1.3017794  1.3223126   12 1.3848343
[12,]    13    1.3728775  1.3895116   13 1.4569254
[13,]    14    1.4439755  1.4572791   14 1.5290164
[14,]    15    1.5150736  1.5255895   15 1.6011075

[[1]]$VP:
[[1]]$VP$cc:
[1] 3.78138

[[1]]$VP$mat:
         matx    mat.ret   vec.ret     matx vecx
 [1,]     2    0.5148744  0.5953111    2 0.5803748
 [2,]     3    0.5859725  0.6589516    3 0.6524659
 [3,]     4    0.6570705  0.7227423    4 0.7245569
 [4,]     5    0.7281686  0.7867244    5 0.7966480
 [5,]     6    0.7992667  0.8509471    6 0.8687391
 [6,]     7    0.8703648  0.9154666    7 0.9408301
```

```
[7,]       8 0.9414628 0.9803449    8 1.0129212
[8,]       9 1.0125609 1.0456453    9 1.0850123
[9,]      10 1.0836590 1.1114273   10 1.1571033
[10,]     11 1.1547570 1.1777402   11 1.2291944
[11,]     12 1.2258551 1.2446159   12 1.3012854
[12,]     13 1.2969532 1.3120648   13 1.3733765
[13,]     14 1.3680513 1.3800736   14 1.4454676
[14,]     15 1.4391493 1.4486078   15 1.5175586

[[1]]$PL:
[[1]]$PL$cc:
[1] 9.711457

[[1]]$PL$mat:
      matx    mat.ret    vec.ret    matx vecx
[1,]     2  0.5322204  0.6585367    2 0.5918326
[2,]     3  0.6033184  0.7217828    3 0.6639237
[3,]     4  0.6744165  0.7850594    4 0.7360148
[4,]     5  0.7455146  0.8483762    5 0.8081058
[5,]     6  0.8166126  0.9117461    6 0.8801969
[6,]     7  0.8877107  0.9751859    7 0.9522880
[7,]     8  0.9588088  1.0387170    8 1.0243790
[8,]     9  1.0299069  1.1023669    9 1.0964701
[9,]    10  1.1010049  1.1661696   10 1.1685611
[10,]   11  1.1721030  1.2301669   11 1.2406522
[11,]   12  1.2432011  1.2944085   12 1.3127433
[12,]   13  1.3142991  1.3589513   13 1.3848343
[13,]   14  1.3853972  1.4238571   14 1.4569254
[14,]   15  1.4564953  1.4891895   15 1.5290164

[[2]]:
[[2]]$BHT:
[[2]]$BHT$cc:
[1] NA

[[2]]$BHT$mat:
      matx    mat.ret    vec.ret    matx vecx
[1,]     2  0.6443588  0.6751473    2 0.7327161
[2,]     3  0.7154569  0.7411473    3 0.8048071
[3,]     4  0.7865550  0.8076939    4 0.8768982
[4,]     5  0.8576530  0.8748106    5 0.9489892
[5,]     6  0.9287511  0.9424978    6 1.0210803
[6,]     7  0.9998492  1.0107327    7 1.0931714
[7,]     8  1.0709472  1.0794733    8 1.1652624
[8,]     9  1.1420453  1.1486638    9 1.2373535
[9,]    10  1.2131434  1.2182418   10 1.3094446
[10,]   11  1.2842415  1.2881442   11 1.3815356
[11,]   12  1.3553395  1.3583119   12 1.4536267
[12,]   13  1.4264376  1.4286923   13 1.5257177
[13,]   14  1.4975357  1.4992406   14 1.5978088
[14,]   15  1.5686337  1.5699198   15 1.6698999

[[2]]$VP:
[[2]]$VP$cc:
[1] NA
```

```
[[2]]$VP$mat:
       matx   mat.ret    vec.ret    matx   vecx
 [1,]     2 0.4925103  0.5188190     2 0.5656183
 [2,]     3 0.5636084  0.5852944     3 0.6377094
 [3,]     4 0.6347064  0.6523382     4 0.7098004
 [4,]     5 0.7058045  0.7199536     5 0.7818915
 [5,]     6 0.7769026  0.7881209     6 0.8539825
 [6,]     7 0.8480006  0.8568001     7 0.9260736
 [7,]     8 0.9190987  0.9259368     8 0.9981647
 [8,]     9 0.9901968  0.9954691     9 1.0702557
 [9,]    10 1.0612948  1.0653337    10 1.1423468
[10,]    11 1.1323929  1.1354707    11 1.2144378
[11,]    12 1.2034910  1.2058267    12 1.2865289
[12,]    13 1.2745891  1.2763559    13 1.3586200
[13,]    14 1.3456871  1.3470203    14 1.4307110
[14,]    15 1.4167852  1.4177892    15 1.5028021

[[2]]$PL:
[[2]]$PL$cc:
[1] 7.424668


[[2]]$PL$mat:
       matx   mat.ret    vec.ret    matx   vecx
 [1,]     2 0.5272022  0.6350115     2 0.5885339
 [2,]     3 0.5983002  0.6983460     3 0.6606250
 [3,]     4 0.6693983  0.7617391     4 0.7327161
 [4,]     5 0.7404964  0.8252093     5 0.8048071
 [5,]     6 0.8115945  0.8887802     6 0.8768982
 [6,]     7 0.8826925  0.9524812     7 0.9489892
 [7,]     8 0.9537906  1.0163494     8 1.0210803
 [8,]     9 1.0248887  1.0804294     9 1.0931714
 [9,]    10 1.0959867  1.1447735    10 1.1652624
[10,]    11 1.1670848  1.2094409    11 1.2373535
[11,]    12 1.2381829  1.2744946    12 1.3094446
[12,]    13 1.3092810  1.3399970    13 1.3815356
[13,]    14 1.3803790  1.4060041    14 1.4536267
[14,]    15 1.4514771  1.4725584    15 1.5257177
```

2.11.2.3 Mechanical Cycling Demarcation Maps

To access the GUI for mechanical cycling demarcation maps, left-click on the "Coffin Manson demarcation map" entry in the demarcation map dropdown menu. This brings up the default GUI for the function "demarc.CM." In Figure 2.17 we show the filled in GUI that produced Figure 2.8. Left-clicking the apply with the mouse results in Figure 2.8, and the output below stored in the vector "dum" in Splus.

```
  dum
$life:
       matx       mat.ret
 [1,] 1e-005  -1.3333333  -0.18911360
 [2,] 1e-004  -1.0000000   0.06088972
 [3,] 1e-003  -0.6666667   0.31089688
```

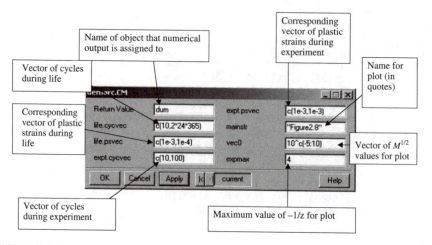

FIGURE 2.17
GUI for Coffin–Manson demarcation map function in Splus for mechanical and thermal cycling, filled in to produce Figure 2.8.

```
 [4,]  1e-002  -0.3333333   0.56091229
 [5,]  1e-001   0.0000000   0.81094548
 [6,]  1e+000   0.3333333   1.06101696
 [7,]  1e+001   0.6666667   1.31117079
 [8,]  1e+002   1.0000000   1.56150148
 [9,]  1e+003   1.3333333   1.81221051
[10,]  1e+004   1.6666667   2.06372253
[11,]  1e+005   2.0000000   2.31691025
[12,]  1e+006   2.3333333   2.57347557
[13,]  1e+007   2.6666667   2.83640022
[14,]  1e+008   3.0000000   3.10991211
[15,]  1e+009   3.3333333   3.39794514
[16,]  1e+010   3.6666667   3.70143618

$expt:
         matx       mat.ret
 [1,]  1e-005  -1.3333333  -0.98620244
 [2,]  1e-004  -1.0000000  -0.65286910
 [3,]  1e-003  -0.6666667  -0.31953577
 [4,]  1e-002  -0.3333333   0.01379756
 [5,]  1e-001   0.0000000   0.34713090
 [6,]  1e+000   0.3333333   0.68046423
 [7,]  1e+001   0.6666667   1.01379756
 [8,]  1e+002   1.0000000   1.34713090
 [9,]  1e+003   1.3333333   1.68046423
[10,]  1e+004   1.6666667   2.01379756
[11,]  1e+005   2.0000000   2.34713090
[12,]  1e+006   2.3333333   2.68046423
[13,]  1e+007   2.6666667   3.01379756
[14,]  1e+008   3.0000000   3.34713090
[15,]  1e+009   3.3333333   3.68046423
[16,]  1e+010   3.6666667   4.01379756

>
```

The first column of both the life and the expt. matrices are the vectors of $M^{1/z}$ values; the remaining columns are the $-1/z$ values computed from the demarcation equations.

3

Interface for Building Kinetic Models

3.1 Description and Concepts behind the Interface

To perform the calculations in Section 2.7, it is necessary to have a way of specifying a kinetic model so the computer can solve the relevant system of differential equations. In the remaining chapters we also need this ability. This chapter is devoted to describing the reasoning we used in creating an interface within which it is relatively easy to specify a kinetic model.

One of the extremely useful aspects of linear models theory in statistics is the ability to design experiments and analyze data in a way that allows comparing many models quickly with the result. A key aspect of this is the ability to easily represent, both conceptually and computationally, whole families of models.

To deal with this, we take advantage of the correspondence noted in Section 1.6.2 of Chapter 1 (second section of the complement). First-order chemical kinetic models have three representations:

They can be represented mathematically as the system of differential equations:

$$\frac{dA_t}{dt} = -k_1 A_t$$

$$\frac{dA_t}{dt} = k_1 A_t$$
(3.1)

or in matrix form, for example, by the equation:

$$\frac{d}{dt}\begin{pmatrix} A_{1t} \\ A_{2t} \end{pmatrix} = \begin{pmatrix} -k_1 & 0 \\ k_1 & 0 \end{pmatrix}\begin{pmatrix} A_{1t} \\ A_{2t} \end{pmatrix}$$
(3.2)

We can also use the following graphical representation for the first-order transmutation of A_1 to A_2.

$$A_1 \xrightarrow{k_1} A_2 \qquad (3.3)$$

Note that we use a single arrow to represent the reaction because it is a unidirectional transformation with a single thermal barrier. We will designate it as a single-arrow process. The latter two forms are the ones most useful in thinking about how to easily generate and communicate representations of first-order kinetic models. It should be recalled that k_1 in the above representations is a function in its own right of the various stresses to which the system is subject.

One problem in attempting to code these relationships by hand is keeping track of the conservation relationships in the matrices like Equation 3.2. The interface provided for Splus provides this tracking automatically.

To describe the interface, suppose we start with two versions of Equations 3.3, and consider these building blocks:

$$\begin{aligned} A_1 \xrightarrow{k_1} A_2 &\approx \frac{\partial}{\partial t}\begin{pmatrix} A_{1t} \\ A_{2t} \end{pmatrix} = \begin{pmatrix} -k_1 & 0 \\ k_1 & 0 \end{pmatrix}\begin{pmatrix} A_{1t} \\ A_{2t} \end{pmatrix}, \\ A_3 \xrightarrow{k_2} A_4 &\approx \frac{\partial}{\partial t}\begin{pmatrix} A_{3t} \\ A_{4t} \end{pmatrix} = \begin{pmatrix} -k_2 & 0 \\ k_2 & 0 \end{pmatrix}\begin{pmatrix} A_{3t} \\ A_{4t} \end{pmatrix} \end{aligned} \qquad (3.4)$$

If we have specified these two building blocks, what can we build?

One possibility is to equate A_2 and A_3. In this case we obtain a sequential process:

$$A_1 \xrightarrow{k_1} A_2 \xrightarrow{k_2} A_4 \approx \frac{\partial}{\partial t}\begin{pmatrix} A_{1t} \\ A_{2t} \\ A_{4t} \end{pmatrix} = \begin{pmatrix} -k_1 & 0 & 0 \\ k_1 & -k_2 & 0 \\ 0 & k_2 & 0 \end{pmatrix}\begin{pmatrix} A_{1t} \\ A_{2t} \\ A_{4t} \end{pmatrix} \qquad (3.5)$$

Alternatively, we can equate A_1 and A_3 in Equation 3.4, in which case we obtain a competing process (competing for the resources of A_1):

$$A_1 \begin{array}{c} \xrightarrow{k_2} A_4 \\ \xrightarrow{k_1} A_2 \end{array} \approx \frac{\partial}{\partial t}\begin{pmatrix} A_{1t} \\ A_{2t} \\ A_{4t} \end{pmatrix} = \begin{pmatrix} -(k_1+k_2) & 0 & 0 \\ k_1 & 0 & 0 \\ k_2 & 0 & 0 \end{pmatrix}\begin{pmatrix} A_{1t} \\ A_{2t} \\ A_{4t} \end{pmatrix} \qquad (3.6)$$

We can both equate A_1 and A_3 and equate A_2 and A_4 in Equation 3.4, in which case we have a process in which there are two separate paths from A_1 to A_2 (a rejoining process):

$$A_1 \begin{array}{c} \xrightarrow{k_1} \\ \xrightarrow{k_2} \end{array} A_2 \approx \frac{\partial}{\partial t}\begin{pmatrix} A_{1t} \\ A_{2t} \end{pmatrix} = \begin{pmatrix} -(k_1+k_2) & 0 \\ (k_1+k_2) & 0 \end{pmatrix}\begin{pmatrix} A_{1t} \\ A_{2t} \end{pmatrix} \qquad (3.7)$$

Interface for Building Kinetic Models

We can equate A_1 and A_4 and equate A_2 and A_3 in Equation 3.4, in which case we have a process in which there is a cycle (a reversible process):

$$A_1 \underset{k_2}{\overset{k_1}{\rightleftarrows}} A_2 \approx \frac{\partial}{\partial t}\begin{pmatrix} A_{1t} \\ A_{2t} \end{pmatrix} = \begin{pmatrix} -k_1 & k_2 \\ k_1 & -k_2 \end{pmatrix}\begin{pmatrix} A_{1t} \\ A_{2t} \end{pmatrix} \quad (3.8)$$

Finally, we can just equate A_2 and A_4 in Equation 3.4, in which case we mix the two sources A_1 and A_3, in the same final state:

$$\begin{matrix} A_1 \xrightarrow{k_1} \\ A_3 \xrightarrow{k_2} \end{matrix} A_4 \approx \frac{\partial}{\partial t}\begin{pmatrix} A_{1t} \\ A_{3t} \\ A_{4t} \end{pmatrix} = \begin{pmatrix} -k_1 & 0 & 0 \\ 0 & -k_2 & 0 \\ k_1 & k_2 & 0 \end{pmatrix}\begin{pmatrix} A_{1t} \\ A_{3t} \\ A_{4t} \end{pmatrix} \quad (3.9)$$

Exercise 3.1.1

[Mathematical exercise] Demonstrate that the above five possibilities are the only distinct ways to use two distinct arrows between as many as three distinct states.

The above five processes show an approach to taking two simple processes and combining them into a more complex process. The approach is based on equating beginning and/or end states in one process with beginning and/or end states in the second process. Interestingly, it is fairly easy to program the matrix modification in such a way that when beginning and end states are defined, these five processes can combine two arbitrarily complex first-order kinetic systems.

This approach supplies a fairly fast way to assemble new kinetic systems; unfortunately, as mentioned in Complement 1.6.2, it is not a complete construction kit. One of the simplest kinetic models that cannot be constructed using this approach has four states with every state connected reversibly with every other state. Substituting the numbers 1 to 4 for the states A_1 to A_4, the directed graph or arrow diagram for this process is shown in Figure 3.1.

Exercise 3.1.2

The mathematically inclined reader should prove that it is impossible to create Figure 3.1 using the above five processes. (*Hint:* The trick is to use the fact that only beginning and ending states can be used in defining new processes.)

To reach this kinetic system, we define a new operation, which "draws" a single arrow between any two states. Clearly, this single operation plus the ability to define a number of states would be enough to create any arrow

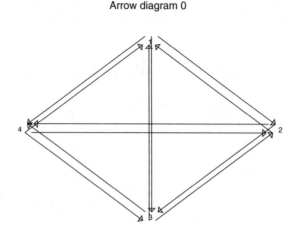

FIGURE 3.1
Simplest process that cannot be generated using the five two-arrow processes as generating schema.

diagram. However, this approach requires more work on the part of the analyst than does using the five assembly steps above and inserting specific arrows only when necessary.

Some complex kinetic systems are useful to think of as primitive. For example, the migration process in Section 2.7 is an example of a special kind of one-dimensional diffusion process. The simplest diffusion process is Fickian diffusion. In one dimension this has the form:

$$A_1 \underset{k_b}{\overset{k_f}{\rightleftarrows}} A_2 \underset{k_b}{\overset{k_f}{\rightleftarrows}} \cdots \underset{k_b}{\overset{k_f}{\rightleftarrows}} A_n \qquad (3.10)$$

where $k_f = k_b$.

Another useful primitive system is glassy relaxation. Glasses and polymers both have significant randomness in their structure. Therefore, it is reasonable to suppose that the local energy surface has some randomness. In glasses, for example, high energy, in the form of two ultraviolet photons, or a single gamma photon, may cause the charges to separate, forming an "exciton." The charges can be trapped apart by the shape of the energy surface. Because the surface is random, the activation energy to recombine is random as well. This can be represented roughly by a model of the form shown in Figure 3.2. This model is specified as a primitive model, the central state is the native glass, the many outer states represent the different configurations. Experimental evidence to date suggests that in at least some glasses, the activation energies are random but the premultiplier is constant.

In addition to specifying the form of the directed graph or arrow diagram corresponding to a model, it is also necessary to identify the way the "rate

Interface for Building Kinetic Models 69

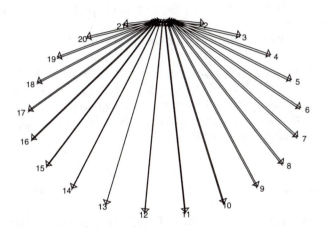

FIGURE 3.2
Arrow diagram for glassy relaxation process with 20 jointly observable states.

constants" corresponding to each arrow vary with the environment. Thus, in first-order chemical kinetics, although the system of differential equations at a given stress may be represented by a matrix, it is actually necessary for a function to be built to track the stress dependencies. The easy way to do this is to write a function that creates the matrix, which then goes to the differential equation solver.

Occasionally, for physical reasons the user may wish to specify some common physical parameters among some of the arrows. Thus, the combination programs using the operations described in Equations 3.5 through 3.9, as well as internally inserting arrows, must be able to keep track of names. This makes using an object-oriented programming language, in which the "objects" can include both functions and identifiers of various types, almost a necessity. Because the language also must be able to automatically create functions in itself, it has to have a self-referential structure such as LISP.

For actually solving the system of differential equations fast enough so that they can be solved across a computational demarcation map, or recursively fit to data, a set of points with equal time intervals can be interpolated into the data or the time interval over which we want to check for a maximum. Then solving the system at a given stress simply involves the solution of the matrix exponential at a single time followed by repeated multiplication of the state vector by the matrix exponential. This can be done quickly enough so that optimization algorithms such as the Powell or modified Gauss–Newton fits based on numerical differentiation can be run in minutes to hours to days on a laptop PC, depending on the complexity of the system and the number of data sets. Bates and Watts (1988) discuss an alternative approach based on taking the eigenvalues of the rate constant matrix, and on taking semisymbolic derivatives. The reader is welcome to experiment with different approaches and improve on this.

Mathematical note: Generalizing the approach presented in this book to nonlinear systems of differential equations, which is necessary for a program to automatically handle higher-order chemical kinetics, requires a very fast, very stable, nonlinear differential equation solver. We have not been able to find such yet, although an operator exponentiation scheme looks promising. An alternative approach would be to somehow simplify the optimization so the equations have to be solved only once in a neighborhood of the parameter space, and some of the optimization can be done using approximations without fully solving the system.

3.2 Complement to Chapter 3: Our Interface in Splus, Kinetic Data Objects, and the GUIs to Create Them

Creating a kinetic model on the computer is a fairly simple process. The approach is to create a collection of component models, and then combine them as needed. Because some higher-order processes have arisen repeatedly in our work, there are a number of component models built to reflect those higher-order models directly.

3.2.1 Creating Components of the Kinetic Model

To access a given GUI for a component model, left-click on "Kinetic model creation" above the toolbar menu, and then left-click on "create a component of the process" on the resulting drop-down menu. This gives a second drop-down menu with components:

1. Discrete diffusion
2. Creep
3. Stress voiding
4. Onestep
5. Onestep with second variable
6. Onestep with stress
7. Onestep with second variable, and stress
8. Glassy onestep

Below we list each of these in sequence, describe the GUIs brought up by left-clicking on them, and describe the result of applying them.

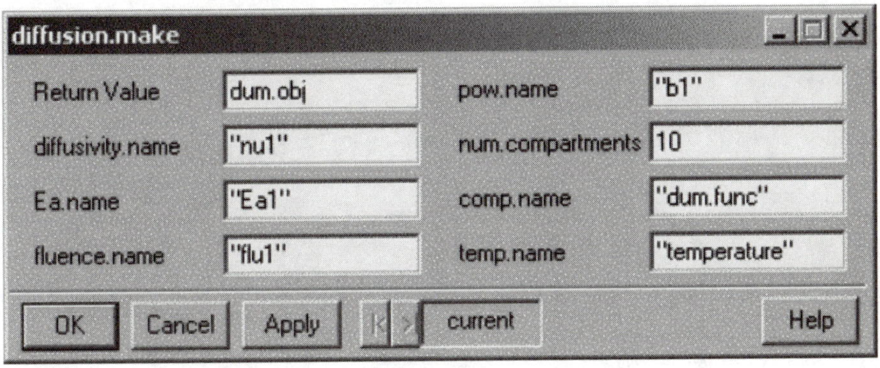

Interface for Building Kinetic Models

diffusion.make			
Return Value	dum.obj	pow.name	"b1"
diffusivity.name	"nu1"	num.compartments	10
Ea.name	"Ea1"	comp.name	"dum.func"
fluence.name	"flu1"	temp.name	"temperature"
OK Cancel Apply		< > current	Help

FIGURE 3.3
Filled-in GUI for discrete diffusion process.

3.2.1.1 Discrete Diffusion

Left-clicking on this entry brings up Figure 3.3, which we have filled in. The model object this creates has two components, the dynamic structure, corresponding to the arrow diagram, and the rate constant. The rate constant has the mathematical form:

$$(\text{flu})^{b1}(10^{nu1})\exp\left(-\frac{E_{a1}}{k(\text{temperature}+273)}\right)$$

Here temperature is assumed to be on the centigrade scale. The rate constant for this is the same for all arrows. The arrow diagram specified by the GUI in Figure 3.3 is shown in Figure 3.4 with arrows going both ways between each compartment. This is a simple one-dimensional diffusion model. Even with this few compartments, the distribution of something starting in compartment 1 starts looking very much like a classical diffusion process. Note that the axes mean nothing; the important part is the directed graph formed by the arrows and the numbered vertices in the plot.

The object that is produced by applying this function has the form:

```
> dum.obj
$parameter.vec:
[1] "nu1" "Ea1" "b1"
```
(this component of the object contains the names of all the parameters)

```
$envir.vec:
[1] "temperature" "flu1"
```
(this component of the object contains the names of the environmental variables)

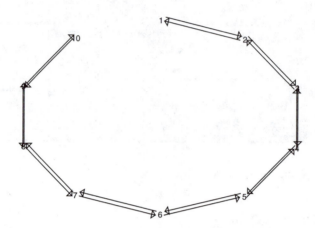

FIGURE 3.4
Arrow diagram for discrete diffusion process created from GUI in Figure 3.3.

```
$func:
function(parvec, envvec)
{
        nu1 <- parvec[1]
        Ea1 <- parvec[2]
        b1 <- parvec[3]
        temperature <- envvec[1]
        flu1 <- envvec[2]
        T1 <- temperature + 273
        z1 <- flu1^b1 * (10^nu1) * exp(- Ea1/(boltzmann * T1))
        z1 * diffmat.skell.closed(10)
}
```
(this component of the object contains the actual function just created)

```
$comp.name:
[1] "dum.func"
```
(this component of the object contains the names of the function just created)

```
$follow.vec:
  [1] "dum.func" "dum.func" "dum.func" "dum.func" "dum.func" "dum.func"
  [7] "dum.func" "dum.func" "dum.func" "dum.func"
```
(this component of the object associates the name of the function just created with the vector of states; it is useful in keeping track of what states correspond to what final function in composite functions)

```
$workspace.vec:
[1] "dum.func" "diffmat.skell.closed"
```
(this component carries the function names for transfer into a dump file, so composite functions contain all necessary names for transfer to another system with the package on it)

Interface for Building Kinetic Models 73

It specifies not only the function, but also names for environmental effects, names for physical parameters, and various other labels useful for the combination programs in helping keep track of how different components have been put together.

3.2.1.2 Creep

Creep is a self-diffusion process driven by a gradient, enhanced by temperature. It has the same sort of arrow diagram, but has separate forward and backward rate processes. The forward rate process has the activation energy barrier reduced by a function of the mechanical stress; the backward rate process has the activation energy barrier increased by the same amount. Figure 3.5 is the GUI (filled in) that comes up when we left-click on "creep." We have filled in the values ahead of time. The rate constants take the form:

$$K_{forward} = (10^{nu1})\exp\left(-\frac{(E_{a1} - (\text{alphaXmech.stress}))_+}{k(\text{temperature} + 273)}\right)$$

$$K_{backward} = (10^{nu1})\exp\left(-\frac{(E_{a1} + (\text{alphaXmech.stress}))}{k(\text{temperature} + 273)}\right)$$

Here the $(a)_+$ indicates the positive branch is evaluated. The arrow diagram corresponding to applying the above GUI is given in Figure 3.6.

3.2.1.2.1 Viewing the Arrow Diagrams

The arrow diagrams are generated by left-clicking on "visualize arrow diagrams" on the first drop-down menu from the kinetic model creation heading. The GUI to create the arrow diagram shown in Figure 3.6 is shown in Figure 3.7.

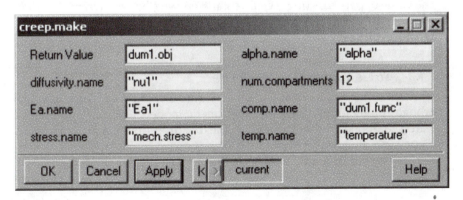

FIGURE 3.5
Filled-in GUI for creep process.

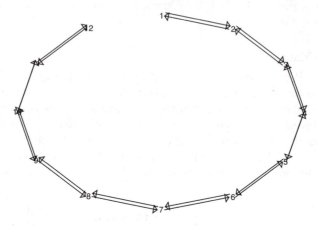

FIGURE 3.6
Arrow diagram for creep process created from the GUI in Figure 3.5.

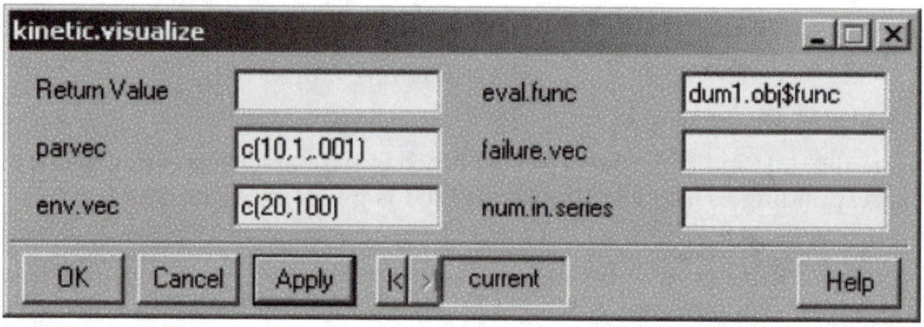

FIGURE 3.7
Filled-in GUI to visualize arrow diagram for creep process.

The value of parvec is $(\log_{10}(v), E_a, \alpha)$, and env.vec, has some sample values for temperature and mechanical stress. The entry for eval.func is just the part of the dum1.obj structure with the function.

3.2.1.3 Stress Voiding

The GUI for stress voiding is shown as Figure 3.8, filled in. The blank GUI is brought up by left-clicking first on "kinetic model creation," and then on "create a component of the process" in the first drop-down menu, and finally on "stress voiding."

The mathematical model corresponding to the GUI has the form:

$$A_1 \underset{k_b}{\overset{k_f}{\rightleftarrows}} A_2 \underset{k_b}{\overset{k_f}{\rightleftarrows}} \cdots \underset{k_b}{\overset{k_f}{\rightleftarrows}} A_n \qquad (3.11)$$

Interface for Building Kinetic Models

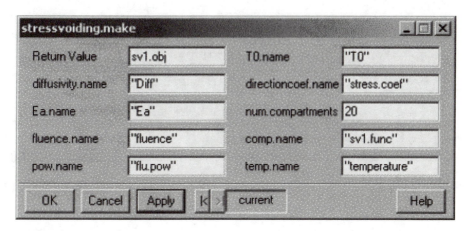

FIGURE 3.8
Filled-in GUI for stress voiding.

where

$$k_f = (\text{fluence})^{\text{flu.pow}} 10^{\text{Diff}} \exp\left(-\frac{E_a - \text{stress.coef} \times (T0 - \text{temperature})_+}{k(\text{temperature} + 273)}\right) \quad (3.12)$$

and

$$k_b = (\text{fluence})^{\text{flu.pow}} 10^{\text{Diff}} \exp\left(-\frac{E_a + \text{stress.coef} \times (T0 - \text{temperature})_+}{k(\text{temperature} + 273)}\right). \quad (3.13)$$

Clicking the apply button results in the generation of the object "sv1.obj" in Splus.

The arrow diagram is predictably just like that for creep and diffusion, only with 20 states. Notice that above, T_0 is included as a parameter. The reason we chose to do this is that T_0, the temperature at which stress becomes 0 on the material system in question, may be difficult to know *a priori* unless very careful attention is paid to the temperature profile followed in manufacture. However, we do use it as a potentially adjustable parameter in experiment design, which with extreme care is at least conceptually possible.

3.2.1.4 One-Step Process

The GUI for the one-step process is shown in Figure 3.9, filled in to create the function "onestepa0.func." The blank GUI is brought up by left-clicking first on "kinetic model creation," and then on "create a component of the process" in the first drop-down menu, and finally on "Onestep."

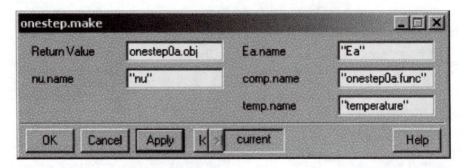

FIGURE 3.9
Filled-in GUI for one-step Arrhenius process.

FIGURE 3.10
Arrow diagram for one-step Arrhenius process.

This arrow diagram is the simplest possible, one arrow between two states shown in Figure 3.10. The mathematical form is

$$A_1 \xrightarrow{k_1} A_2 \text{ or alternatively } \frac{d}{dt}\begin{pmatrix} A_1 \\ A_2 \end{pmatrix} = \begin{pmatrix} -k_1 & 0 \\ k_1 & 0 \end{pmatrix}\begin{pmatrix} A_1 \\ A_2 \end{pmatrix}$$

with

$$k_1 = 10^{nu} \exp\left(-\frac{Ea}{k(\text{temperature}+273)}\right).$$

Interface for Building Kinetic Models

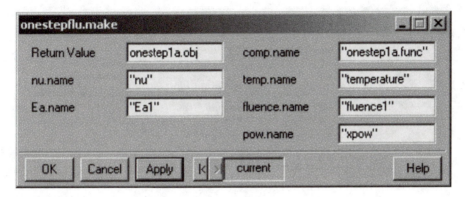

FIGURE 3.11
Filled-in GUI for one-step process, combined Arrhenius, and power-law dependence.

3.2.1.5 One Step with Second Variable

This function allows a second "stress" to be added to the Arrhenius model in the form of a power law, premultiplying the Arrhenius. The mathematical form is now

$$A_1 \xrightarrow{k_1} A_2 \text{ or alternatively } \frac{d}{dt}\begin{pmatrix} A_1 \\ A_2 \end{pmatrix} = \begin{pmatrix} -k_1 & 0 \\ k_1 & 0 \end{pmatrix}\begin{pmatrix} A_1 \\ A_2 \end{pmatrix}$$

with

$$k_1 = \text{fluence1}^{xpow} 10^{nu} \exp\left(-\frac{Ea}{k(\text{temperature}+273)}\right)$$

based on the way the GUI is filled out in Figure 3.11. The blank GUI is brought up by left-clicking first on "kinetic model creation," and then on "create a component of the process" in the first drop-down menu, and finally on "Onestep with second variable."

3.2.1.6 One Step with Stress

This function allows a "stress" to be added to the Arrhenius model in the form of a modification of the activation energy. This sort of model can arise either through mechanical stress (e.g., Krauss and Erying, 1975) or through an indirect band gap effect (e.g., Kittel and Kroemer, 1980). The mathematical form is now

$$A_1 \xrightarrow{k_1} A_2 \text{ or alternatively } \frac{d}{dt}\begin{pmatrix} A_1 \\ A_2 \end{pmatrix} = \begin{pmatrix} -k_1 & 0 \\ k_1 & 0 \end{pmatrix}\begin{pmatrix} A_1 \\ A_2 \end{pmatrix}$$

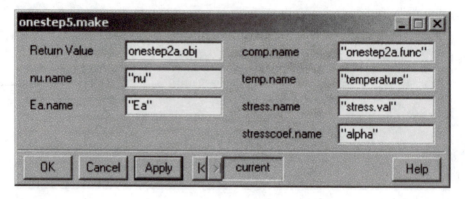

FIGURE 3.12
Filled-in GUI for one-step process, combined Arrhenius, and mechanical stress dependence.

with

$$k_1 = 10^{nu} \exp\left(-\frac{(Ea - alpha * stress.val)_+}{k(temperature + 273)}\right)$$

Here $(x)_+$ represents the positive part of x, so it is 0 if x is negative. In the case of an indirect band gap effect the stress.variable is the energy in a single photon at the wavelength in question. The alpha premultiplier is then constrained to be between 0 and 1 by the conservation of energy.

The GUI corresponding to the formula above is filled out in Figure 3.12. The blank GUI is brought up by left-clicking first on "kinetic model creation," and then on "create a component of the process" in the first drop-down menu, and finally on "Onestep with stress."

3.2.1.7 One Step with Second Variable and Stress

This function combines the two models above. The mathematical form is now

$$A_1 \xrightarrow{k_1} A_2 \text{ or alternatively } \frac{d}{dt}\begin{pmatrix}A_1\\A_2\end{pmatrix} = \begin{pmatrix}-k_1 & 0\\k_1 & 0\end{pmatrix}\begin{pmatrix}A_1\\A_2\end{pmatrix}$$

with

$$k_1 = fluence2^{ypow} 10^{nu} \exp\left(-\frac{(Ea - alpha * stress.val)_+}{k(temperature + 273)}\right)$$

Interface for Building Kinetic Models

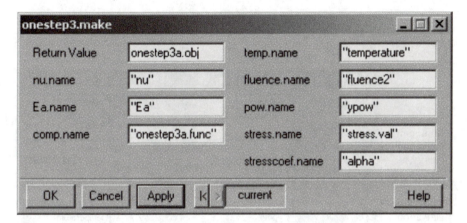

FIGURE 3.13
Filled-in GUI for one-step process, combined Arrhenius, mechanical stress dependence, and power law dependence.

The GUI corresponding to the formula above is filled out in Figure 3.13. The blank GUI is brought up by left-clicking first on "kinetic model creation," and then on "create a component of the process" in the first drop-down menu, and finally on "One step with second variable and stress."

3.2.1.8 Glassy One Step

This function is used in particular to denote the formation of defects with a distribution of activation energies through the action of some stress for which a power law is a good model. Formation of defects in amorphous silicon, or in silica, through exposure to high-energy photons is one example of such a process. The relaxation then comes about through thermal relaxation. Built into this model is the notion that there is a distribution of activation energies. The mathematical form is now

$$A_1 \underset{k_{2i}}{\overset{k_1}{\rightleftarrows}} \begin{pmatrix} A_2 \\ \cdot \\ \cdot \\ \cdot \\ A_n \end{pmatrix} \text{ or alternatively } \frac{d}{dt} \begin{pmatrix} A_1 \\ \cdot \\ \cdot \\ \cdot \\ A_n \end{pmatrix} = \begin{pmatrix} -k_1 & k_{21} & \cdot & \cdot & k_{2n} \\ k_1 p_2 & -k_{21} & \cdot & \cdot & 0 \\ \cdot & 0 & \cdot & \cdot & \cdot \\ \cdot & \cdot & \cdot & \cdot & 0 \\ k_1 p_n & 0 & \cdot & \cdot & -k_{2n} \end{pmatrix} \begin{pmatrix} A_1 \\ \cdot \\ \cdot \\ \cdot \\ A_n \end{pmatrix}$$

with

$$k_1 = \text{fluence3}^{\text{ypow}} \text{con1}$$

$$k_{2i} = 10^{\text{nu}} \exp\left(-\frac{E_{ai}}{kT}\right)$$

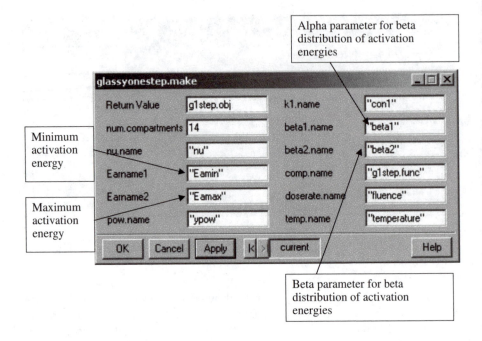

FIGURE 3.14
Filled-in GUI for glassy relaxation process.

The sum

$$\sum_{i=2}^{n} p_i = 1$$

is chosen to be a discrete approximation to a beta distribution scaled between a minimum and a maximum activation energy. Thus, the distribution of activation energies is specified with four parameters, the two parameters for the beta distribution and the minimum and maximum activation energy.

The GUI corresponding to the formula above is filled out in Figure 3.14. The blank GUI is brought up by left-clicking first on "kinetic model creation," and then on "create a component of the process" in the first drop-down menu, and finally on "Glassy onestep." The arrow diagram now has the form shown in Figure 3.15.

There are 14 states beyond the native state without defects. The beta distribution was not chosen for any physical reason. It was chosen simply because it provides a flexible, easily parameterized way of describing a distribution of activation energies.

Interface for Building Kinetic Models

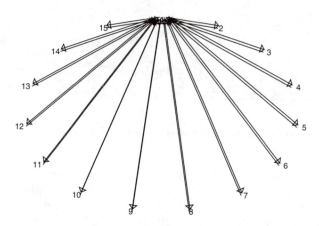

FIGURE 3.15
Arrow diagram for glassy relaxation process created by applying the GUI in Figure 3.14.

3.2.2 Combining Submodels

The software to combine the models is based on the mathematical patterns described in the complement to Chapter 1, and in Section 3.1. To access a given GUI for combining models using one of these operations, left-click on "Kinetic model creation" above the toolbar menu, and then left-click on "create a higher level process" on the resulting drop-down menu. From there, the six operations described in Section 3.1 are available. The list of operations is as follows:

1. Competing reaction
2. Mixing reaction
3. Reversible reaction
4. Rejoining reaction
5. Sequential reaction
6. Simple connection of internal states

We step through using the component processes created above to show how to combine given processes.

3.2.2.1 Competing Reactions

The competing reaction operation combines any two processes at their beginning, creating two branches from a single source, competing for that source. The GUI for the competing process operation comes up by left-clicking on

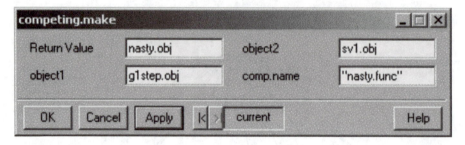

FIGURE 3.16
Filled-in GUI for combining glassy relaxation process with stress voiding process using the competing process schema.

the "competing reactions" entry. For illustration purposes, we show how the competing process works with two complex processes, the glassy relaxation and the stress voiding process. The glassy relaxation object was stored as "g1step.obj," and the stress voiding object was stored as "sv1.obj." To combine the two in a competing reaction type of process we fill out the competing reaction GUI as in Figure 3.16 and apply it.

The Splus object, nasty.object looks like:

```
> nasty.obj
$parameter.vec:
 [1] "nu"         "Eamin"         "Eamax"       "ypow"
 [5] "con1"       "beta1"         "beta2"       "Diff"
 [9] "Ea"         "flu.pow"       "T0"          "stress.coef"

$envir.vec:
[1] "fluence"     "temperature"

$func:
function(p3, e3)
{
        parvec1 <- p3[c(c(1, 2, 3, 4, 5, 6, 7))]
        parvec2 <- p3[c(c(8, 9, 10, 11, 12))]
        env1 <- e3[c(1, 2)]
        env2 <- e3[c(2, 1)]
        m1 <- g1step.func(parvec1, env1)
        m2 <- sv1.func(parvec2, env2)
        matcomp(m1, m2)
}

$comp.name:
[1] "nasty.func"

$follow.vec:
 [1] "g1step.func"  "g1step.func"  "g1step.func"  "g1step.func"
 [5] "g1step.func"  "g1step.func"  "g1step.func"  "g1step.func"
 [9] "g1step.func"  "g1step.func"  "g1step.func"  "g1step.func"
[13] "g1step.func"  "g1step.func"  "g1step.func"  "sv1.func"
```

Interface for Building Kinetic Models

```
[17]  "sv1.func"     "sv1.func"     "sv1.func"     "sv1.func"
[21]  "sv1.func"     "sv1.func"     "sv1.func"     "sv1.func"
[25]  "sv1.func"     "sv1.func"     "sv1.func"     "sv1.func"
[29]  "sv1.func"     "sv1.func"     "sv1.func"     "sv1.func"
[33]  "sv1.func"     "sv1.func"

$workspace.vec:
[1]  "nasty.func"         "Eadistmatbeta"         "g1step.func"
[4]  "sv1.func"           "migmat.skell.closed"   "matcomp"
```

Notice that there are now 34 states and two environmental factors: fluence and temperature. Both the stress voiding and the glassy relaxation process had fluence variables, and they might have been different. However, because the same string was used to denote them, they have been combined in the new model. *This is important to note. It is very important to pay attention to naming conventions, because common names are interpreted by the program to have common meanings.* This holds for both environments and parameters.

Assuming the fluences were a common stress, we can proceed. The parameter vector in the object tells us how many, and what parameters we need to supply. The arrow diagram for this process looks like Figure 3.17.

The parameter vector input to obtain this picture had the values: c(5,.1,1,1.1,.00001,1,1,6,.4,1.2,150,.0001), with the parameters in the order given above.

This example serves to show that the approach works with complex processes; however, it is easiest to understand with simpler processes. Thus, for example, we create a second one-step process with different activation energy and nu value as in Figure 3.18.

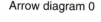

Arrow diagram 0

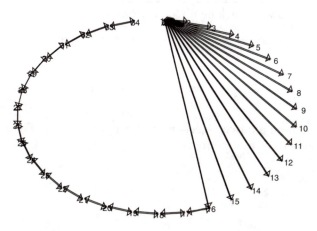

FIGURE 3.17
Arrow diagram created by applying the GUI in Figure 3.16.

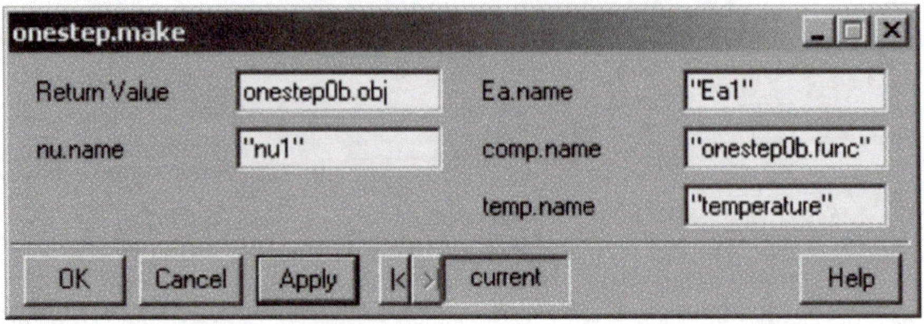

FIGURE 3.18
Filled-in GUI for creating a second simple, one-step process with an Arrhenius temperature dependence and independent parameters.

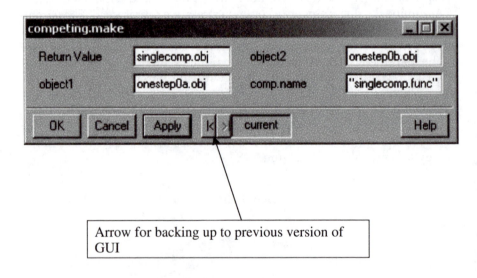

FIGURE 3.19
Filled-in GUI for combining two one-step processes using the competing process schema.

We can combine the two objects with this operator as in Figure 3.19. The object this generates has the form:

```
> simplecomp.obj
$parameter.vec:
[1] "nu"  "Ea"  "nu1"  "Ea1"

$envir.vec:
[1] "temperature"
```

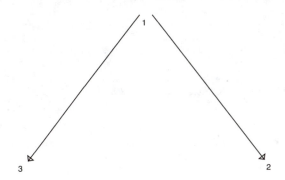

FIGURE 3.20
Arrow diagram resulting from applying the GUI in Figure 3.19.

```
$func:
function(p3, e3)
{
        parvec1 <- p3[c(c(1, 2))]
        parvec2 <- p3[c(c(3, 4))]
        env1 <- e3[1]
        env2 <- e3[1]
        m1 <- onestep0a.func(parvec1, env1)
        m2 <- onestep0b.func(parvec2, env2)
        matcomp(m1, m2)
}

$comp.name:
[1] "simplecomp.func"

$follow.vec:
[1] "onestep0a.func"  "onestep0a.func"  "onestep0b.func"

$workspace.vec:
[1] "simplecomp.func"  "onestep0a.func"   "onestep0b.func"
[4] "matcomp"
```

The resulting arrow diagram is shown in Figure 3.20.

3.2.2.2 Mixing Reactions

The mixing reaction operation combines any two processes at their ends, creating two branches to a single product, mixing in that product. The GUI for the mixing process operation comes up by left-clicking on the "mixing reactions" entry. The GUI filled out to combine the two simple one-step

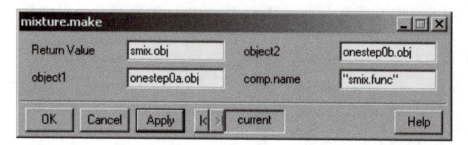

FIGURE 3.21
Filled-in GUI for combining two one-step processes using the mixture process schema.

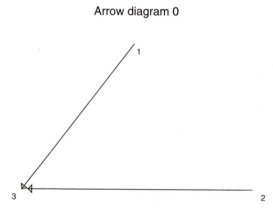

FIGURE 3.22
Arrow diagram resulting from applying the GUI in Figure 3.21.

processes into a simple mixing reaction has the form shown in Figure 3.21. The arrow diagram has the form shown in Figure 3.22.

3.2.2.3 Reversible Reactions

The reversible reaction operation identifies the beginning of one process with the end of the other, and visa versa, creating a cycle in the directed graph. The GUI for the reversible process operation comes up by left-clicking on the "reversible reactions" entry. The GUI filled out to combine the two simple one-step processes into a simple reversible reaction is shown in Figure 3.23, and the arrow diagram in Figure 3.24.

Interface for Building Kinetic Models 87

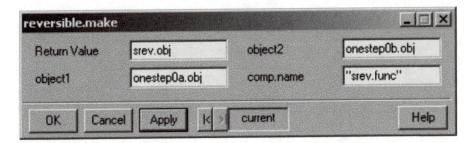

FIGURE 3.23
Filled-in GUI for combining two one-step processes using the reversible or equilibrating process schema.

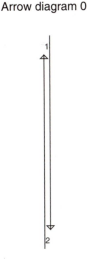

FIGURE 3.24
Arrow diagram resulting from applying the GUI in Figure 3.23.

3.2.2.4 Rejoining Reactions

The rejoining reaction operation identifies the beginnings and ends of two processes with one another The GUI for the rejoining process operation comes up by left-clicking on the "rejoining reactions" entry. Filling out the GUI to rejoin two simple processes will result in just a single arrow, because there are only two states; however, the arrow will represent the sum of the two rates from the two different processes. To actually show something different we combine the competing process with the reversible process in the GUI (Figure 3.25), and show the resulting object and arrow diagram (Figure 3.26).

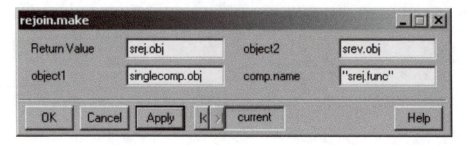

FIGURE 3.25
Filled-in GUI for combining competing and equilibrium processes using the rejoining process schema.

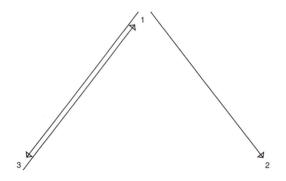

FIGURE 3.26
Arrow diagram resulting from applying the GUI in Figure 3.25.

```
> srej.obj
$parameter.vec:
[1] "nu"   "Ea"   "nu1" "Ea1"

$envir.vec:
[1] "temperature"

$func:
function(p3, e3)
{
        parvec1 <- p3[c(c(1, 2, 3, 4))]
```

Interface for Building Kinetic Models

```
        parvec2 <- p3[c(c(1, 2, 3, 4))]
        env1 <- e3[1]
        env2 <- e3[1]
        m1 <- simplecomp.func(parvec1, env1)
        m2 <- srev.func(parvec2, env2)
        matrej(m1, m2)
}

$comp.name:
[1] "srej.func"

$follow.vec:
[1] "onestep0a.func" "onestep0a.func" "onestep0b.func" "onestep0b.func"

$workspace.vec:
[1] "srej.func"        "simplecomp.func" "onestep0a.func"
[4] "onestep0b.func"   "matcomp"         "srev.func"
[7] "matcyc"           "matrej"
```

Notice because of the way we created them from the same simple, one-step functions, there are four rather than eight parameters. Figure 3.26 is still deceptive because the arrow leading from state 1 to 3 is actually the sum of two rates. The extra complexity then shows up in the stress dependence of the rates, rather than in the fundamental dynamics given by the arrow diagram.

3.2.2.5 Sequential Reactions

The sequential reaction operation identifies the beginning of a second process with the end of a first, creating a sequential pathway. The GUI for the sequential process operation comes up by left-clicking on the "sequential reactions" entry. The GUI filled out to combine the two simple, one-step processes into a simple sequential reaction is shown in Figure 3.27, and the corresponding arrow diagram in Figure 3.28.

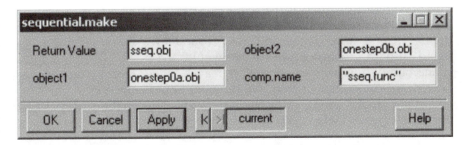

FIGURE 3.27
Filled-in GUI for combining two one-step processes using the sequential process schema.

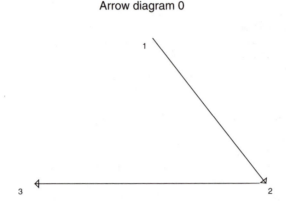

FIGURE 3.28
Arrow diagram resulting from applying the GUI in Figure 3.27.

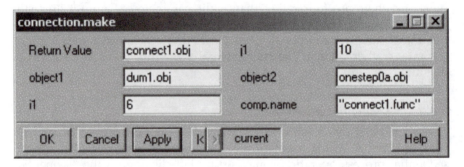

FIGURE 3.29
Filled-in GUI for connecting states 6 with 10 of a previously constructed creep process (Figures 3.5 and 3.6) using a single arrow.

3.2.2.6 *Simple Connection of Internal States*

This operator takes any two states in a process, and connects them using a one-step, or one-step reversible process. So, for example, if we have the creep process created earlier (shown in Figure 3.6) and if we wish to connect states 6 and 10 with a one-step process we would access the GUI for "simple connection of internal states" on the "create a higher level process" of the "kinetic model creation" menu, and fill it out as in Figure 3.29.

Clicking "apply" produces the object shown below with the corresponding arrow diagram in Figure 3.30.

```
> connect1.obj
$parameter.vec:
[1] "nu1"    "Ea1"    "alpha"  "nu"     "Ea"
```

Interface for Building Kinetic Models

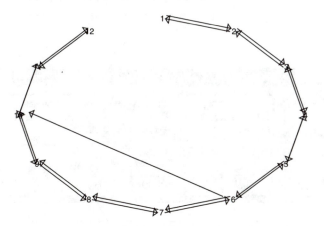

FIGURE 3.30
Arrow diagram resulting from applying the GUI in Figure 3.29.

```
$envir.vec:
[1] "temperature" "mech.stress"

$func:
function(p3, e3)
{
        parvec1 <- p3[c(1:3)]
        parvec2 <- p3[c(c(4, 5))]
        env1 <- e3[c(1, 2)]
        env2 <- e3[1]
        m1 <- dum1.func(parvec1, env1)
        m2 <- onestep0a.func(parvec2, env2)
        matconnect(m1, 6, 10, m2)
}

$comp.name:
[1] "connect1.func"

$follow.vec:
 [1] "dum1.func" "dum1.func" "dum1.func" "dum1.func" "dum1.func"
 [6] "dum1.func" "dum1.func" "dum1.func" "dum1.func" "dum1.func"
[11] "dum1.func" "dum1.func"

$workspace.vec:
[1] "connect1.func"      "dum1.func"         "migmat.skell.closed"
[4] "onestep0a.func"     "matconnect"
```

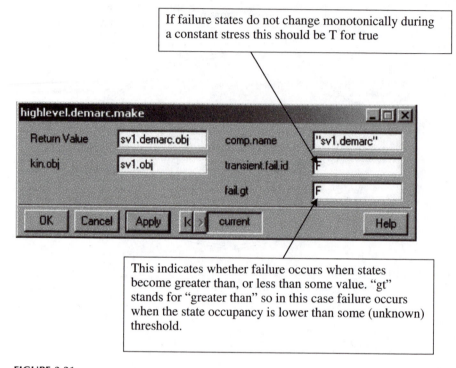

FIGURE 3.31
Filled-in GUI to create a demarcation mapping function from the stress voiding kinetic model.

3.2.3 Computational Demarcation Map Example from Chapter 2

In the appendix of the book, when installing the software, the reader is told to create a folder in the object explorer containing the functions. In this section we show one of the reasons for taking advantage of that instruction.

Left-click on the demarcation maps drop-down menu, and then on "create function for direct kinetic evaluation using kinetic model." The kinetic function for the example in Section 2.7 is sv1.obj created in Section 3.2.1.3. The filled-in GUI is shown in Figure 3.31.

Applying this GUI immediately creates the function "sv1.demarc." The function can now be found in the function folder created in the Splus objects explorer. Left-clicking on "sv1.demarc" in the functions folder in the objects explorer brings up the GUI shown in Figure 3.32 filled in to calculate the demarcation matrix for the life stress.

In parnames the vector of parameter names is input in the same order it appears in sv1.demarc.object. The parameter vectors are listed below:

```
> nuvec.chp2
[1]  -4   0   4   8  12  16  20
> eavec.chp2
[1] 0.3 0.6 0.9 1.2 1.5
```

Interface for Building Kinetic Models 93

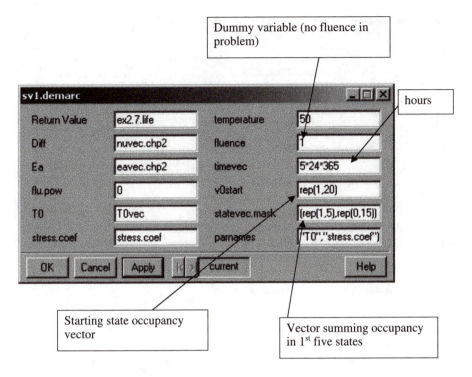

FIGURE 3.32
Filled-in GUI for the stress-voiding demarcation mapping function, creating the reaction extent matrix for life.

```
> T0vec
[1] 200 350 500
> stress.coefvec
[1] 1e-006 1e-005 1e-004 1e-003 1e-002
>
```

Recall that vectors are created by using the function c(). So nuvec.chp2<-c(-4,0,4,8,12,16,20) is the command for creating the vector nuvec.chp2.

The GUI filled in to create the matrix for the comparison experiment is shown in Figure 3.33. While running, the GUI displays the word "working" in the upper banner. Once that is run it is possible to create the display shown as Figure 2.11. Left-click on the command "plot extent matrices from the direct kinetic eval" under the "demarcation maps" drop-down menu. We show the GUI in Figure 3.34 filled in to calculate the display in Figure 2.11. To determine the way to fill out the columns it is helpful to print out the first few lines of one of the matrices, e.g.,

```
> ex2.7.life[1:5,]
      Diff  Ea flu.pow  T0 stress.coef       extent
[1,]   -4 0.3       0 200       1e-006 4.996426e+000
```

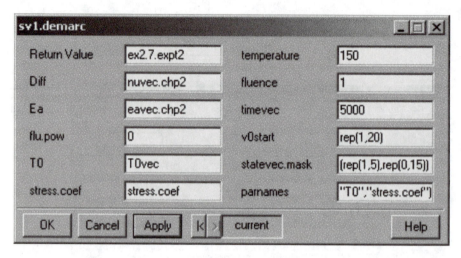

FIGURE 3.33
Filled-in GUI for the stress-voiding demarcation mapping function, creating the reaction extent matrix for experiment 2.

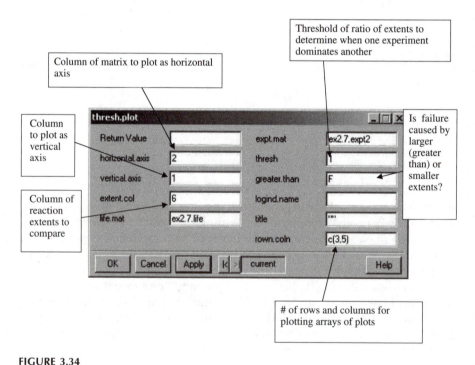

FIGURE 3.34
Filled-in GUI for plotting the comparison of the life reaction extent matrix with the experiment 2 reaction extent matrix.

Interface for Building Kinetic Models 95

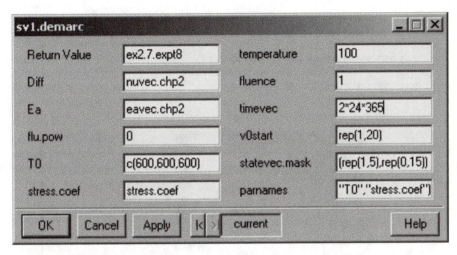

FIGURE 3.35
Filled-in GUI for the stress voiding demarcation mapping function, creating the reaction extent matrix for experiment 8.

```
[2,]   -4 0.3     0 200     1e-005 4.964240e+000
[3,]   -4 0.3     0 200     1e-004 4.625068e+000
[4,]   -4 0.3     0 200     1e-003 4.952939e-026
[5,]   -4 0.3     0 200     1e-002 0.000000e+000
```

We see that the activation energies (the horizontal axis on the plots) occupy the second column. The diffusivities (the vertical axis) are the first. The extent column is the sixth column. The value in thresh is very important. Here we assume that the same makeup in the material has the same effect on failure at the two different temperatures. If, in fact, the problem caused is expected to be a short caused by increased conductivity, and at 50°C, the effect of degradation is half that at 150°C, this number could be ½ instead of 1.

Left-clicking the apply button then recreates Figure 2.11. Also in the report window, the total number of points, tot, and the number of points where life dominates, NL, are supplied.

In Figure 3.35 we show the GUI to create the two-year 100°C experiment with T_0 at 600°C. Note, the T_0 value is entered as a three-place vector c(600,600,600), to ensure that the matrices are the same size.

We leave it as an exercise for the reader to produce the final two experiments in the table. To create plot 2.13, it is necessary to access the final functionality on the demarcation map drop-down menu. Left-click on "plot INT of extent matrices from direct kinetic eval" on the menu. In our analysis, we assigned the matrices for experiments 11 and 12 the names "ex2.7.expt11" and "ex2.7.expt12" using the return value slot in the GUI for sv1.demarc. To create the appropriate data structure for the new GUI, type:

```
> ex2.7.11and12<-list(ex2.7.expt11,ex2.7.expt12)
```

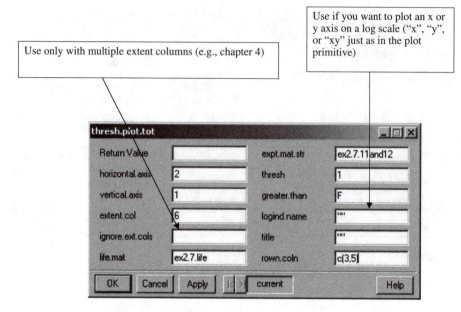

FIGURE 3.36
Filled-in GUI for plotting the comparison of life with two experiments (11 and 12) simultaneously.

and hit the carriage return. The GUI filled in correctly to produce Figure 2.13 is shown as Figure 3.36.

Exercise 3.2.1

Further exploration and verification of the results in the example are left as an exercise for the reader.

4
Evanescent Process Mapping

In Chapter 2, we began with the notion that the information on reliability in an accelerated test for devices meant to have ultrahigh reliability is in the failure- and degradation-free periods prior to any observable degradation or failure. To extract this information, we developed the notion of demarcation mapping. However, the analytical demarcation maps restrict consideration to single-step processes, while computational demarcation maps restrict consideration to just a single chosen alternative process.

The purpose of this chapter is to take our analysis of null result accelerated tests one step farther. As a start, we assume that we have done some tests on the device in question, observed failures or some kind of degradation, and found the most parsimonious physical model that fits the data (model with the fewest estimated parameters), and that model predicts high reliability for the device. Further, we assume that this model is a first-order kinetic model. (The development of a theory of evanescent process mapping for nonlinear kinetic models is a topic of current research, and so beyond the scope of this presentation).

Certainly, a minimal requirement for a model that is used for extrapolating a set of data is that it fit that data. However, given two points on a plane, there is one straight line and an infinite number of parabola and hyperbola that fit those two points. Clearly, if we are extrapolating, it is useful to be able to rule out the parabola and hyperbola.

Working in the world of first-order kinetic models, if a model is the simplest one that fits a data set, how can we characterize a set of slightly more complex first-order kinetic models that also fit that data set, and result in significantly different extrapolations? Further, how can we reduce the risk that these models are true experimentally?

In this chapter we begin with theoretical development. We start by showing that a natural neighborhood to consider for first-order kinetic models is a neighborhood characterized by weighted directed graphs close to the one corresponding to the model. We go on to define a notion of risk orthogonality over an experiment. The notion of *evanescent* processes, processes that vanish over the original experiments but become evident in other parts of domain space, is defined by considering those processes that are risk orthogonal to the original processes. A least favorable model (one that is close to the original, but will definitely result in failure at operating conditions, and will not be detectable at highly accelerated conditions) is then defined, and a theorem shows how to construct one as an evanescent process. The smallest neighborhood of models guaranteed to contain a least favorable model is the space we sample for evaluating how to reduce risk. Finally, we describe how the risk orthogonality helps us in applying the extrapolation theorem from Chapter 2.

In the next section we give examples of neighborhoods of particular models, and discuss the notion of statistical sampling from these neighborhoods to estimate risk. In particular we advocate a conservative approach.

In the third section we give an example considering how to distribute samples among multiple experiments. In the fourth section we describe what the theory presented here implies about the capability and limitations of accelerated testing.

4.1 Building Blocks for the Theory

4.1.1 Model Neighborhoods

In Complement 1.6.2 we noted that a first-order kinetic process has three mathematically equivalent representations:

It can be represented mathematically as the system of differential equations:

$$\frac{dA_t}{dt} = -k_1 A_t$$

$$\frac{dB_t}{dt} = k_1 A_t$$

or in matrix form by the equation:

$$\frac{d}{dt}\begin{pmatrix} A_t \\ B_t \end{pmatrix} = \begin{pmatrix} -k_1 & 0 \\ k_1 & 0 \end{pmatrix} \begin{pmatrix} A_t \\ B_t \end{pmatrix}$$

We can also use the following graphical representation for the first-order transmutation of A to B:

$$A \xrightarrow{k_1} B \tag{4.1}$$

It is the third representation we find useful here, both because the mathematics of weighted directed graphs is quite simple and manageable and because the graphical representation provides some physical intuition.

Evanescent Process Mapping

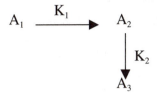

FIGURE 4.1
Arrow diagram for sequential process leading to failure.

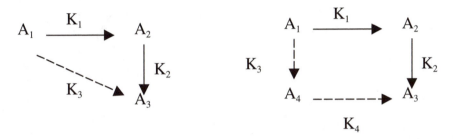

FIGURE 4.2
Arrow diagrams for two evanescent processes associated with the process shown in Figure 4.1.

Now suppose that based on a series of exacting, but very accelerated experiments, we deduce that over the regime we have observed it, a failure mode is well described by the kinetic process shown in Figure 4.1, where A_3 is the failure causing product of the process. Suppose, for example, the transition from A_1 to A_2 is caused by mechanical shocks to a case, causing microcracks that allow moisture ingress. Suppose the transition from A_2 to A_3 is caused by corrosion of components inside the case. Two potential evanescent processes are shown in Figure 4.2.

In the first, we suppose that there is a direct method for moisture to ingress and cause corrosion at low stress levels, perhaps a slow diffusion process (here we take slight liberty with the notation). Alternatively, we could suppose that a different way for damage to occur to the casing may exist. For example, if the casing is plastic, exposure to ultraviolet and/or ozone can lower the susceptibility to cracking so much that ordinary stresses might cause the cracking.

To consider a neighborhood of the original model, we keep the original "weight" functions associated with the original graph. Then one way of defining a neighborhood is to consider the following:

1. All unweighted graphs that can be created by adding up to m points and up to n arrows to the original unweighted graph.
2. For each such unweighted graph, all possible weighted graphs with the original weights (possibly range of weights specified by confidence

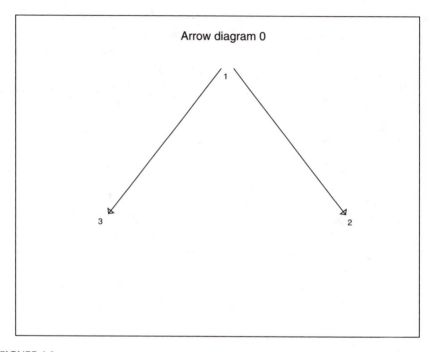

FIGURE 4.3
Arrow diagram for unobservable masking, or unobservable censoring schema as a worst-case evanescent process.

bounds) on the original graph (now subgraph) and then all other possible weight functions on the arrows in the remaining subgraph.

This defines a simple neighborhood of weighted graphs (and hence system of linear differential equations consistent with first-order kinetics) with the parameters m and n defining a function that substitutes for distance. The first question we need to consider is "How big a neighborhood do we want to consider?"

There is a simple kind of kinetic model that provides the worst-case scenario for the use of accelerated testing. That is a kinetic model, which at lower stress drives the flaws to failure, but at higher stress drives the flaws into some kind of passivated state (Figure 4.3, state 2 is failure, state 3 is passivation). We call such a model an unobservable masking (or censoring) model, because the passivation masks the failure-causing state without leaving an observable effect. Setting the minimal criteria for realistically assessing risk as including such a model, Theorem 4.1 below defines the neighborhood of models we wish to explore.

We note before stating the theorem, that with two additional arrows, we can add between zero and two new states if we connect to a base model,

M0, and either two or three states if we do not. We restrict ourselves to a maximum of two new arrows and two new states. We confine this theorem to the case where degradation is modeled simply as the accumulation of flaws in a given state. We do not believe this restriction is necessary, but it greatly simplifies the proof. The following results were first given elsewhere (LuValle et al., 2002).

THEOREM 4.1
Start with any arrow diagram-based model that describes degradation or failure at accelerated conditions, and that extrapolates to no failure under operating conditions. The set of all distinguishable arrow diagrams including up to two more arrows and two more states is sufficient to represent an unobservable masking model, for acceleration models at least as complex as the Arrhenius model. (We call the alternative processes in these models evanescent processes.)

PROOF
Consider the original model. Each state corresponding to degradation or failure in the original model is either an absorbing state (no arrows leading from it) or a transient state (at least one arrow leads from it to a state not contributing to degradation). Note that for there to be no failures ever at operating conditions, under this scenario at least one absorbing state in the original model must be passivating (does not cause failure).

Suppose first that at least one degradation or failure state in the original model is absorbing. Then suppose we add two new states (states 1 and 2 in Figure 4.3) and suppose state 3 is the absorbing state in the original model, while states 1 and 2 do not correspond to degradation or failure. Then this model has the potential to be an unobservable censoring model for acceleration parameters at least as complicated as the Arrhenius model.

Suppose that no degradation state in the original model is absorbing. Then add states 1 and 2 and the arrows as before, but let state 3 be a passivation state in the original model (at least at operating conditions). Let state 2 be a degradation state. Again, this model has the potential to be an unobservable censoring model for acceleration parameters at least as complicated as the Arrhenius model.

Since this exhausts the possibilities, the result is proven.

So what do we do with this result? We would like to use it to design experiments to explore the models in this neighborhood. To proceed we need two more building blocks. The first provides some guidance to what sort of experiments we might wish to focus on. The second is some enumeration of these additional models in the neighborhood. Then this can be tied with the theory of Chapter 2 in the calculation of risk over all the additional models.

4.1.2 Risk Orthogonality

We proceed by looking at the kind of experiments that we should focus on. We assume that we have done enough initial experiments at very high stress to define a model, M0, and to show that it will not result in failure during operating life. From the data, for any trajectory through stress space, over time, we can define a time t_0 (along that stress trajectory) such that before that time, we should not see degradation from the process corresponding to our model above the threshold of experimental error. (Or the probability of such is deemed sufficiently low that we need not worry if we believe our model to be true.) For any model M1, which produces change in that early time period, define that effect of that model as a data component for M1, which is risk orthogonal to M0. From this definition we immediately have:

LEMMA 4.2
Any model M1, which causes significant degradation or failure during operating life, has data components that are risk orthogonal to M0.

PROOF
By definition, all operating life is prior to any significant degradation caused by M0.

In practice, we define experimental stress trajectories that give significant time before t_0, but we will probably continue the experiments after t_0 and look for statistically significant differences between the predictions of M0 and the data.

The kinds of physical experiments that allow us to study the risk orthogonal components of such models include low stress experiments, as well as accelerated experiments where stress is stepped down.

4.1.3 Model Enumeration

Exact enumeration of distinct models is difficult; of greater concern is an appropriate algorithm for walking through the distinct models. However, to size the calculation, we can use the following simple upper bound. First we need some definitions: For a given model M0 we define $o_S(M0)$ as the number of states in M0 and $o_A(M0)$ as the number of arrows in M0. Then we have

THEOREM 4.3
Define

$$Z = \left[2 \binom{o_S(M0) + 2}{2} - o_A(M0) \right]$$

where

$$\binom{n}{x} = \frac{n!}{x!(n-x)!}$$

is the number of ways of choosing x states out of n. Then the number of evanescent models in the neighborhood defined in Theorem 4.1 is bounded above by $4[Z(Z-1)] + 2Z$.

PROOF
First, note that the total number of states possible in any of the evanescent models is $o_S(M0) + 2$. Then the number of ways of choosing states to draw an arrow between is

$$\binom{o_S(M0)+2}{2}$$

Because arrows can go either way, the total number of ways to draw one arrow without overwriting an existing arrow is Z. Drawing two arrows without overwriting can be done in $Z(Z-1)$ ways. To obtain the factor of 4, we note that there are four ways we can classify the two new states, PP, PF, FP, and FF, where F is a failure-causing state, and P is a nonfailure-causing state. With only one arrow we can only add one new state, so the premultiplier becomes 2.

We note that this grossly overestimates the number of models. For example, consider the three graphs (actually schema of arrow diagrams) below:

$$M0 \rightarrow F$$

$$M0 \rightarrow F \rightarrow F$$

$$M0 \rightarrow F \leftarrow F$$

If our observable is only the sum of accumulation of flaws in the F states, all three of the above graphs will give identical observable behavior. As our goal is eventually to compute the models, an exact enumeration, while mathematically interesting, is less useful than defining an algorithm for proceeding through arrow diagrams, which identifies behaviorally distinct ones. If our upper bound is small enough, then we can simply walk through all of the diagrams listed from the upper bound, and eliminate those that are behaviorally equivalent as we come on them. To help identify such, we would use a list of forbidden transitions. For example, one forbidden transition motivated by the above is from one failure state that is otherwise absorbing to an absorbing failure state. We return to a more important question. Does Theorem 4.3 provide a small enough bound so that we can completely explore all these possibilities? We can start with the simple one-step model. There are two states, and one arrow in M0, thus Z is

$$\left[2\binom{2+2}{2} - o_A(M0)\right] = 11,$$

and the final number is 462, which is actually quite manageable with appropriate model trimming algorithms to trim out forbidden transitions. On the other hand, consider a model with 34 states and 101 arrows. The final number there is 5,370,806 models to check through. Because checking each model (as shown in the next section) is a fair computational burden (approximately 20 min per experimental stress trajectory on a 1 GHz laptop), this would seem beyond the realm of reasonable calculation without very high power algorithms or tools. In situations like this, it may be possible to find simpler approximate models that would reduce the computational complexity.

4.1.4 Integrating the Theory

We have a base model M0, which predicts no (or very small) degradation or failure during operating life (or during a premaintenance period, etc.) and which fits the observed degradation at operating life. On a computational demarcation map based on the structure of M0, the model M0 is a point (if desired, a region can be built corresponding to uncertainty). Any experiment, which on the computational demarcation map does not reach the uncertainty region about M0, is risk orthogonal to M0.

Using the theory above, we then specify the neighborhood of models around M0, and choose a sample of models from the neighborhood. Ideally, computations would be over the whole neighborhood, but this can be quite a feat. So a carefully specified statistical sample over the neighborhood is what can be used if one is restricted to a laptop using the freeware.

The criterion for an experiment to be a candidate for use under the statistical theory developed in Chapter 2 is that the experiment is risk orthogonal to M0 (so the experiment would cause "minimal" degradation under M0). The best experiments result in the lowest posterior probability for alternative models if it gives a NULL result.

In the next section we show some different examples of neighborhoods, and discuss sampling in a conservative manner from the neighborhood. In the final section we present a nontrivial example of the application of the full theory.

4.2 Identifying Neighborhoods of Models, Sampling, and "Chunking"

There are several ways to identify precisely the models in the neighborhood. The one we have chosen is a simple brute-force approach. We create a list of models, first containing only M0, and then generate the models in Theorem 4.3 one at a time, and compare them to each model in the list. If the generated model is new, it is added to the list. If it is a permutation of an existing model we move on to the next. The current implementation of the selection process is not perfect. We illustrate this with the case where we start with the null process (a single point with no arrow) and enumerate the set of diagrams around it.

From Theorem 4.3,

$$z = \left(2\binom{1+2}{2} - 0\right) = 6$$

so the upper bound is $4[6(5)] + 12 = 132$ possible models. The set of models produced using the current enumeration software in the neighborhood of the NULL model is shown in Figure 4.4.

The long labels denote the degradation states. Note that in some cases the degradation state is being depleted rather than augmented. In some cases, this is the sensible physical way to think about the process, for example, in the formation of a void in an aluminum line. If, for example, we think only of processes where degradation involves adding (something) to a state to cause degradation, then counting from left to right, top to bottom, the diagrams of interest are 1, 3, 5, 6, 7, and 12. These evanescent processes of the NULL process have a special place. Because any degradation process can be considered both the observed process and a NULL process, it is conservative to include these processes as possible evanescent processes.

Suppose that our actual M0 has the form of an equilibrating process (diagram 5 in Figure 4.4). Then from Theorem 4.3,

$$z = \left(2\binom{2+2}{2} - 2\right) = 10$$

so the upper bound is $4[10(9)] + 20 = 380$ possible models. The software enumerates 75 different models from those 380 possible models, shown in Figure 4.5 through Figure 4.7.

Again, review of some of these models may show that they make no sense to consider for the process (e.g., for example, it may make no sense to label

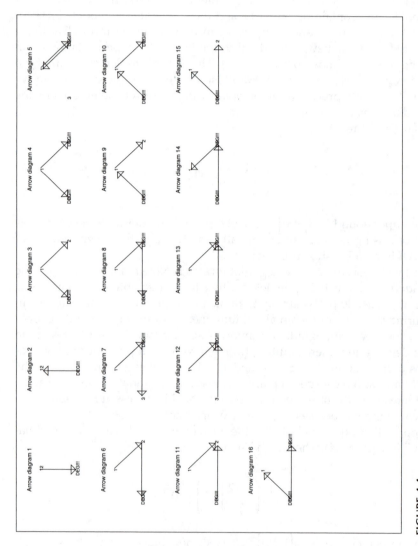

FIGURE 4.4
Neighborhood of arrow diagrams for null model.

Evanescent Process Mapping 107

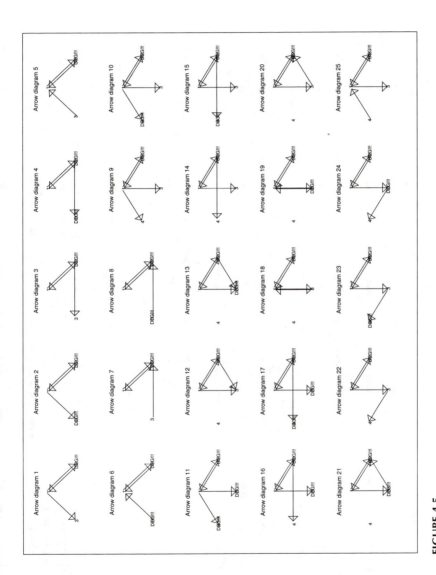

FIGURE 4.5
First 25 unique evanescent processes in neighborhood or reversible model.

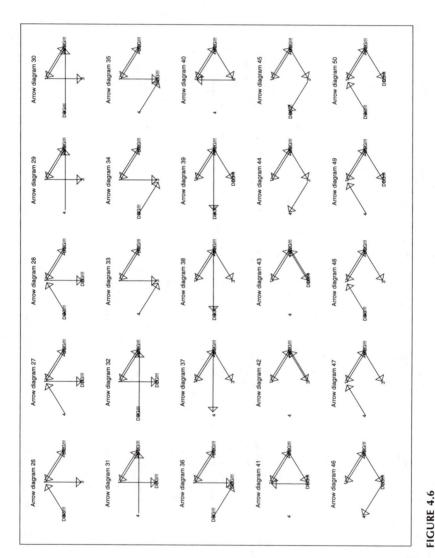

FIGURE 4.6
Second 25 unique evanescent processes in neighborhood or reversible model.

Evanescent Process Mapping 109

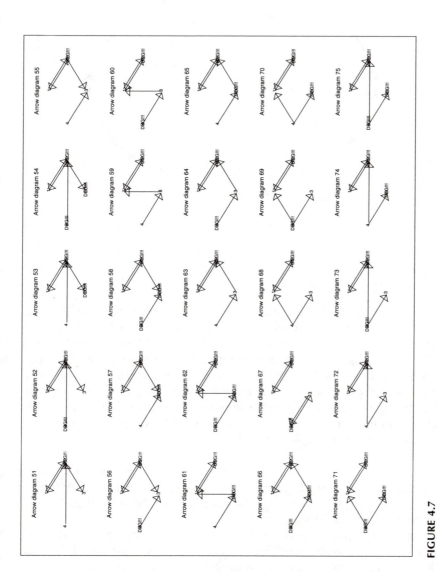

FIGURE 4.7
Last 25 unique evanescent processes in neighborhood or reversible model.

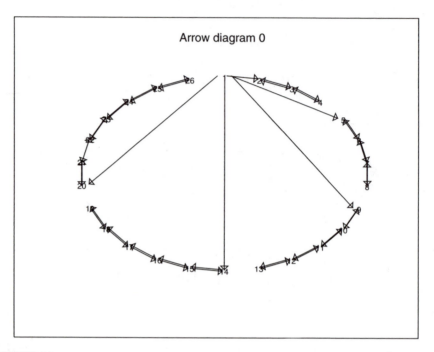

FIGURE 4.8
California hillside with only vertical drainage.

degradation at the root of an arrow for some kinds of processes). Thus, as in the null case, some down selection must be done before sampling.

Once a collection of processes representing the neighborhood is chosen, it is still often necessary to sample from that collection to do the demarcation computations in a reasonable time. The particular approach we have used is to include all processes in the neighborhood of the null model (approximately five depending on the constraints) and then to sample a moderate number from the remaining, perhaps five more. This sample is then used as a basis for approximating the integrals in Equation 2.21, and the implicit integrals in Equation 2.24. The models from the null model neighborhood include at least one potentially least favorable model, so this approach probably provides a reasonable approximation. It is possible that an optimum sampling approach for estimating this integral could be derived, but that is currently an open question.

Sometimes the simple approach above based on the raw graphical representation is not physically sensible. This is particularly true when the kinetic model is used to represent a compartmental system including transport. For example, suppose we consider a series of adjacent columns with a single source, and the columns have different heights. Using our diagrams this might be represented as in Figure 4.8. For example, in a problem estimating probability of a failure of a hillside to support a house in Southern California, this could represent water penetration in adjacent plots of land on a slope.

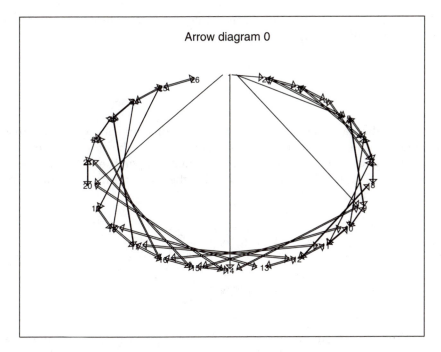

FIGURE 4.9
California hillside with both vertical and horizontal drainage, in the neighborhood of the model in Figure 4.8.

This model, however, is rather simple. Certainly, we would expect adjacent levels of the land to communicate. Such communication is conceptually a very simple addition to the model, but it requires a lot of arrows. Figure 4.9 shows the model with communication.

This example may seem odd in an accelerated testing case, but it is not unreasonable to do testing with scale models, perhaps with appropriately scaled loads to test the effects of moisture and ground cover on the effects of integrity of a hillside under flood conditions.

Clearly, Figure 4.9 should be in the neighborhood of Figure 4.8 from a scientific point of view; however, we need to develop an appropriate way of "chunking" such arrow structures together that works to allow automatic development of neighborhoods. This is another open problem in this area, requiring some kind of combination of topology, algebra, graph theory, and artificial intelligence.

4.3 Example

This example is constructed using a simplified version of Conductive Anodic Filament (CAF) failure mode for printed wiring boards analyzed in Chapter 6. Through the analysis of step stress tests it was possible to show that at lower stress levels, closer to operating conditions, the CAF failure mode had the form of an equilibrating process:

$$A_1 \underset{k_2}{\overset{k_1}{\rightleftarrows}} A_2 \qquad (4.2)$$

Here A_2 is the state that contributes to failure. For simplicity we assume that the rate constants are Arrhenius coefficients. For the sake of the example we assume a device able to withstand very high temperatures. From experiments at 350 and 300°C, our estimates for the temperature dependence of the rate constants are

$$k_1 = 10^{16} \exp\left(-\frac{1.4}{kT}\right)$$
$$k_2 = 10^6 \exp\left(-\frac{0.1}{kT}\right) \qquad (4.3)$$

We assume that failure occurs when A_2 exceeds a threshold, and the data are consistent with that threshold being uniformly distributed between $(1e-3) \times A_{10}$ and A_{10}, where A_{10} is the concentration of A_1 at time 0. We assume that this is mission critical, and we would like to develop an understanding of risk associated with an evanescent process. The normal approach to accelerated testing would be that after identifying the model above, and seeing that no failures occur, in public arguments (e.g., with a customer) we would argue that we looked very hard and see no failure mode that would cause any concern. In private (if we were good at what we did) we would worry very much about evanescent processes, but typically would not have a strategy for checking them.

From experience this works out eight to nine out of ten times; there are not that many stories of hidden failure and degradation modes (although they would tend to be hidden where companies could do so). Thus, we might assign a 10 to 20% chance total to all possible evanescent processes, with a conditional 50% chance that one will actually cause failure given it exists (from our assuming a beta distribution on the probability of observable failure or degradation with $\alpha = 1$, and $\beta = 1$). Thus, the final estimated probability from the exercise will be multiplied by 0.1.

The processes that we could include as part of a neighborhood of the observed degradation mode are listed in Figure 4.4 through Figure 4.7. We have to choose a set of diagrams to represent the actual neighborhood, propose some candidate experiments, set up a prior distribution, and calculate the expected loss for the candidate experiments.

The nature of the reversible process makes it difficult to identify nonsensical evanescent processes automatically, so we will use all 75 listed models

TABLE 4.1

Experiments to Be Numerically Evaluated with Respect to Sampled Evanescent Processes

	Experiment No.						
	1	2	3	4	5	6	7
Time trajectory	1000 h	5000 h	2 yrs	10 h, 4380 h	300 h, 1 yr	1 yr, 300 h	4380 h, 1000 h
Temperature trajectory, °C	200	100	50	200, 50	200, 50	50, 200	50, 200

as the original neighborhood and sample 6 models randomly from those, eliminating models that are noninformative as we go.

For our evanescent processes we can assume either a completely unobserved failure mode (so the process is in the neighborhood of the NULL model corresponding to Figure 4.4) or a complication of the observed process, which results in higher probability of failure (the neighborhood of models corresponding to Figure 4.5 through Figure 4.7).

The models chosen for the neighborhood of the NULL model were models 1, 3, 5, 6, 7, and 12 (numbering from right to left, top to bottom of Figure 4.4). These models represent the possibility of an independent failure mode.

The models from the neighborhood of the estimated reversible process that were chosen by random sampling were models 3, 12, 28, 29, 68, and 70. These models represent possible evanescent extensions of the model, that is, reactions that are part of the process, but do not affect the overall rate significantly at the conditions tested to date. The experiments that were evaluated are listed in Table 4.1.

For the life condition we assume 40°C for 25 years. For each evanescent process model, the free parameters were varied over the following sets:

Activation energy:

$$\left(0, \frac{1}{3}, \frac{2}{3}, 1, \frac{4}{3}, \frac{5}{3}, 2, \frac{7}{3}, \frac{8}{3}, 3\right) \text{ eV}$$

Premultiplier:

$$\left(10^{-6}, 10^{-3}, 10^{0}, 10^{3}, 10^{6}, 10^{9}, 10^{12}, 10^{15}, 10^{18}, 10^{21}\right) \text{ Hz}$$

For any given application it may be possible to restrict the range of either of these parameters either through experience with simple systems or through first principle arguments. An unfortunate aspect of the Arrhenius premultiplier is that there is really no first principle argument for a lower bound on its value, and it truly can vary over many orders of magnitude.

To reach the point where the optimal sample sizes for splitting between the experiments, for reducing the posterior probability can be evaluated, a series of calculations must be stepped through. The sequence of calculations is enumerated below. The way the calculations are grouped corresponds to the way the software works in Splus. The actual commands for this example are illustrated in the first part of the complement of this chapter.

The steps in the analysis are as follows:

1. Select the evanescent processes.
2. Encode the evanescent processes into demarcation map calculators. In the supplied software this is done in a way that groups models with the same number of free and fixed parameters into a single function. This creates a set of affine subspaces in parameter space. Thus for the models selected above:
   - Model 1 from Figure 4.4 has two free parameters, no fixed parameters.
   - Models 3, 5, 6, 7, and 12 from Figure 4.4 have four free parameters and no fixed parameters.
   - Model 3 from Figure 4.5 through Figure 4.7 has two free parameters and four fixed parameters.
   - Models 12, 28, 29, 68, and 70, have four free parameters and four fixed parameters.
3. Evaluate the demarcation criteria over the set of free parameters over each experiment.
4. Combine all of the experiments for each model set, and choose a prior probability measure over the parameter space. The prior measures in the freeware include one measure where each point is equally weighted (denoted hereafter as measure 1) and one where values of the premultiplier below 10^5 Hz are weighted according to the function:

$$\exp\left(-\left(\frac{5-\log_{10}(v)}{2.5}\right)\right)$$

Using each prior, calculate the partition for each experiment for each model, and assess the prior probability for each partition.

5. Guess a starting distribution of samples. Determine a relative weight for each model. Using the theory in Complement 2.11.1, calculate an optimum distribution of samples to minimize the posterior (this distribution changes with the total).

The results for this set of experiments and models are summarized in Table 4.2. The weight assigned to each model is one-quarter for each of the models with two free parameters, and one-half spread evenly among the ten models

TABLE 4.2

Weights for Each Experiment Given Sample Sizes

	Posterior Probability	Experiment No.						
		1	2	3	4	5	6	7
Time trajectory		1000 h	5000 h	2 yrs	10 h, 4380 h	300 h, 1 yr	1 yr, 300 h	4380 h 1000 h
Temperature trajectory, °C		200	100	50	200, 50	200, 50	50, 200	50 200
Measure 1, $N = 70$	0.0319 (0.0327)	3e–3	0.133	0.088	0.043	0.049	0.285	0.399
Measure 1, $N = 1000$	0.0183 (0.0184)	0.029	0.132	0.108	0.062	0.068	0.184	0.417
Measure A, $N = 70$	0.0279 (0.0291)	5e–4	0.121	0.022	0.058	0.044	0.343	0.4115
Measure A, $N = 1000$	0.01455 (0.01463)	0.021	0.123	0.049	0.071	0.062	0.197	0.477

with four free parameters. (Complexity is down-weighted.) α and β are each 1 for the beta distribution. The posterior probability in parentheses is calculated with equal weighting on each experiment. Clearly, the optimum does not buy much. Interestingly, the weights are reminiscent of the kind of weights that would be used to optimize the problem of extrapolating to operating conditions. The conditions closer to operating conditions have higher weights, although the relationship is not perfect.

4.4 Summary, Limitations of Accelerated Testing

When the theory of accelerated testing does not include the information contained in null results (the subject of this chapter and of Chapter 2), the potential problem of evanescent processes is ignored. One consequence is that there is no way to scale effort to desired reliability with a new material system. The effort is always focused on finding a failure mode that we one can

1. prove is artificial, hence ignorable, or
2. bound the effect of, hence use system design to ameliorate.

In the latter case the effort focuses on achieving better bounds for the failure probability or degradation. In the former case there is a fixed expenditure, which "assesses" the reliability.

With evanescent processes admitted as possible (clearly allowed by the laws of physics and chemistry), the Bayesian framework immediately allows us to scale our effort to how much evidence we would like to accumulate against the existence of such processes, up to a hard point, corresponding to the probability assigned to the least favorable models. As those models will be untestable, one thing we have shown in this chapter is that any reliability program for new material systems relying on accelerated testing to prove reliability has unavoidable and irreducible risks.

Thus, if mission critical reliability is essential (e.g., consider the national nuclear storage site at Yucca Mountain, or, extrapolating to much more complicated dynamics, consider the Earth as a system and global warming as a potential degradation process), in addition to the best scientific investigation, it is necessary to provide alternative technology backup systems, based on proven methods. Otherwise, the one implementing the system is accepting risks that are much higher than typically considered acceptable by most of those affected by the decisions.

4.5 Complement to Chapter 4: Using the Evanescent Process Mapping Interface to Duplicate Example 4.3

The evanescent process mapping drop-down menu has five components:

1. Enumerate evanescent processes
2. Select evanescent processes
3. Create demarc array calculator for evanescent processes
4. Identify measure for each partition induced by experiments
5. Optimize nvec to minimize posterior probability for NULL experiment

The example steps through each of these in order. We start by assuming that the model has the form of a reversible or equilibrating process. We are interested in two neighborhoods actually: the neighborhood of $M0$ (the reversible process as listed) and the neighborhood of a NULL model (e.g., a totally independent failure mode, possibly masked statistically by the observed failure mode).

To enumerate the neighborhood of the NULL model, we left-click on the "enumerate evanescent processes" label in the evanescent process drop-down menu to bring up the GUI. To fill out the GUI to evaluate the neighborhood of the NULL model, see Figure 4.10. This will result in production of Figure 4.4 if "apply" is left-clicked.

If we wish instead to enumerate the neighborhood of the reversible model we fill in the GUI as in Figure 4.11. Clicking apply on this results in the three plots corresponding to Figure 4.5 through Figure 4.7.

To create the demarcation maps, for each of the potential evanescent processes we first select the processes. But to select the processes, we need to be able to generate them correctly. All of the models in Figure 4.5 through Figure 4.7 require four one-arrow processes, each with separately specified activation energy and premultiplier. So we create four objects using the one-step GUI. In our case, we labeled them onestep.1.obj, onestep.2.obj,

Evanescent Process Mapping

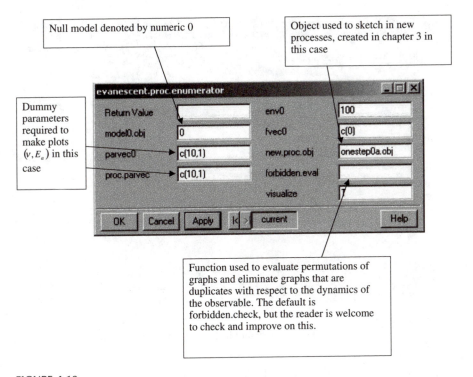

FIGURE 4.10
GUI of evanescent process enumeration function filled out to enumerate in the neighborhood of the NULL model.

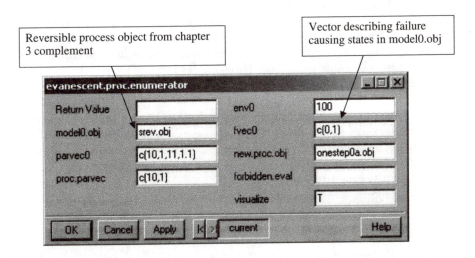

FIGURE 4.11
GUI of evanescent process enumeration function filled out to enumerate in the neighborhood of the simple reversible model produced in the complement to Chapter 3.

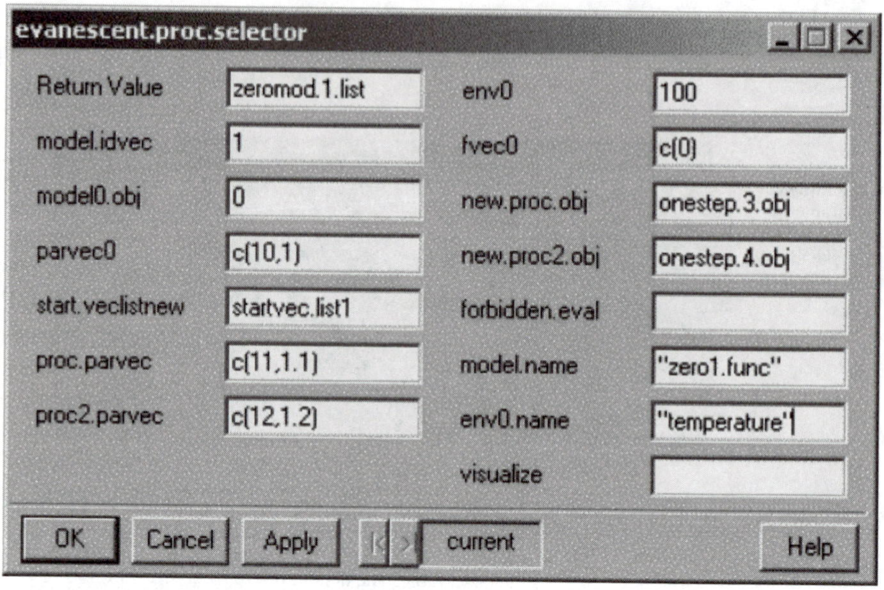

evanescent.proc.selector				_ □ ×
Return Value	zeromod.1.list	env0		100
model.idvec	1	fvec0		c(0)
model0.obj	0	new.proc.obj		onestep.3.obj
parvec0	c(10,1)	new.proc2.obj		onestep.4.obj
start.veclistnew	startvec.list1	forbidden.eval		
proc.parvec	c(11,1,1)	model.name		"zero1.func"
proc2.parvec	c(12,1,2)	env0.name		"temperature"
		visualize		
OK	Cancel	Apply	< > current	Help

FIGURE 4.12
GUI of evanescent process selection function filled out to select model 1 in the NULL model neighborhood to create a list of functions for simultaneous demarcation mapping.

onestep.3.obj, and onestep.4.obj. Because the base model is a reversible model, we create a reversible model based on onestep1.obj and onestep.2.obj.

From Figure 4.4, we chose models 1, 3, 5, 6, 7, and 12. Model 1 requires two free parameters, while models 3, 5, 6, 7, and 12 require four. None of these models includes the original model, so they have no fixed parameters. All of the models from Figure 4.5 through Figure 4.7 include fixed parameters. The demarcation mapping functions work only in given affine subspaces of parameter space. To group models together, the parameters have to belong to the same affine space (even though the system of differential equations may be quite different). Thus, model 1 from Figure 4.4 forms one set, and models 3, 5, 6, 7, and 12 from Figure 4.4 form another set. From Figure 4.5 through Figure 4.7, model 3 forms one set (four fixed parameters, two free), and models 12, 28, 29, 68, and 70 form another (four fixed parameters and four free parameters). To create these four sets of models, we left-click on "select evanescent processes" in the evanescent process map dropdown menu and apply it to the four sets separately. We show each in Figure 4.12 through Figure 4.16.

Notice in Figure 4.12 there is a vector startvec.list1. That is a list of the form:

```
> startvec.list1<-list(c(1,0))
```

This denotes the "concentration" of materials in the two states at the start; for the other models drawn from Figure 4.4 we have

Evanescent Process Mapping

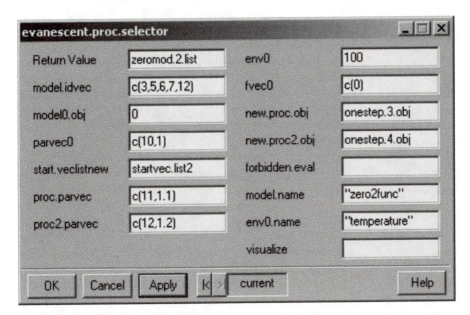

FIGURE 4.13
GUI of evanescent process selection function filled out to select models 3, 5, 6, 7, and 12 in the NULL model neighborhood to create a list of functions for simultaneous demarcation mapping.

```
> startvec.list2<-list(c(1,0,0),c(1,0,0),c(1,0,0),c(1,0,0),c(1,0,1))
```

The GUI for the second group is given in Figure 4.13.

When the GUIs are applied they should also produce a plot of the processes being selected, so for the GUI in Figure 4.13 the plot in Figure 4.14 appears.

The object created by this GUI is a list of model objects. So zeromod.2.list has the form:

```
> zeromod.2.list
[[1]]:
[[1]]$model.obj:
[[1]]$model.obj$parameter.vec:
[1] "nu3" "Ea3" "nu4" "Ea4"

[[1]]$model.obj$envir.vec:
[1] "temperature"

[[1]]$model.obj$func:
function(p3, e3)
{
 parvec1 <- p3[c(1:2)]
 parvec2 <- p3[c(c(3, 4))]
 env1 <- e3[1]
 env2 <- e3[1]
 m1 <- zero2func3a(parvec1, env1)
```

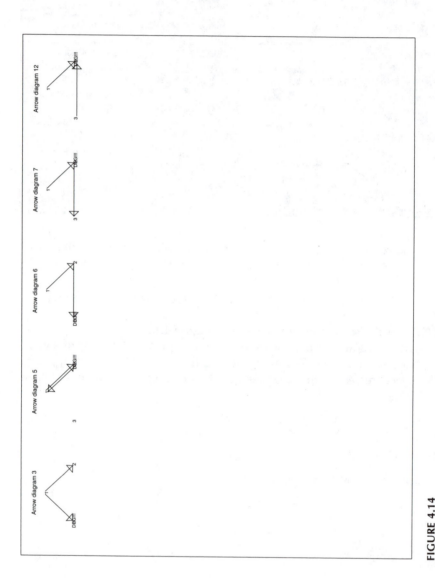

FIGURE 4.14
Plot of the models selected by activating the GUI in Figure 4.13.

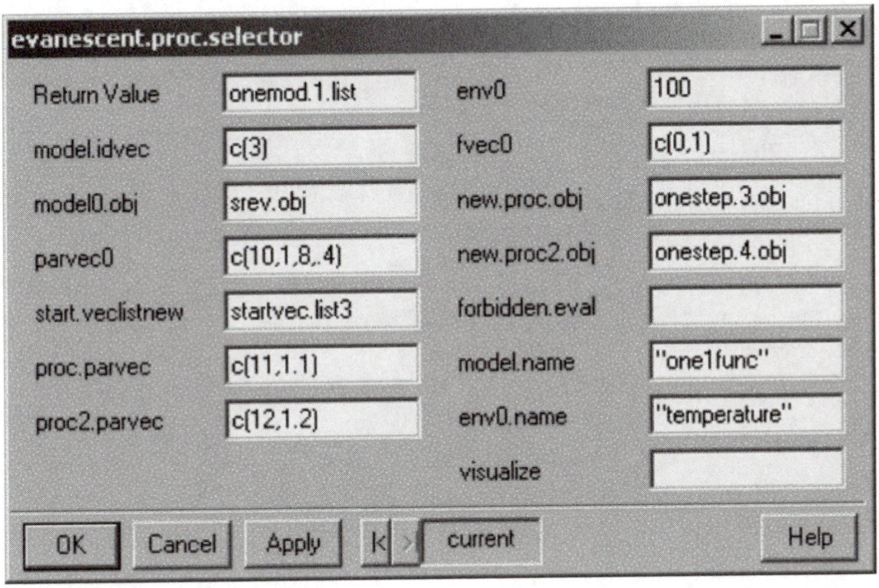

Evanescent Process Mapping

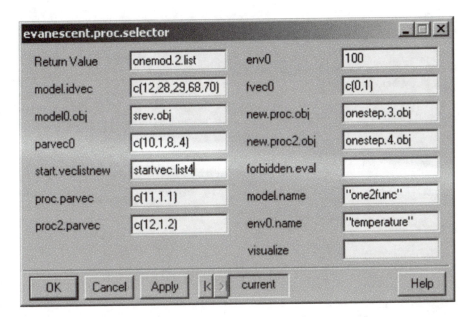

FIGURE 4.15
GUI of evanescent process selection function filled out to select model 3 in the model neighborhood of the simple reversible process to create a list of functions for simultaneous demarcation mapping.

FIGURE 4.16
GUI of evanescent process selection function filled out to select models 12, 28, 29, 68, and 70 in the model neighborhood of the simple reversible process to create a list of functions for simultaneous demarcation mapping.

```
  m2 <- onestep.4.func(parvec2, env2)
  matconnect(m1, 1, 3, m2)
}

[[1]]$model.obj$comp.name:
[1] "zero2func3"

[[1]]$model.obj$follow.vec:
[1] "0" "0"

[[1]]$model.obj$workspace.vec:
[1] "zero2func3"     "zero2func3a"     "zero2func00"     "onestep.3.func"
[5] "matconnect"     "onestep.4.func"

[[1]]$fvec:
[1] 0 0 1

[[1]]$model.id:
[1] 3

[[1]]$transient.id:
[1] F

[[1]]$start.vec:
[1] 1 0 0

[[2]]:
[[2]]$model.obj:
[[2]]$model.obj$parameter.vec:
[1] "nu3" "Ea3" "nu4" "Ea4"

[[2]]$model.obj$envir.vec:
[1] "temperature"

[[2]]$model.obj$func:
function(p3, e3)
{
  parvec1 <- p3[c(1:2)]
  parvec2 <- p3[c(c(3, 4))]
  env1 <- e3[1]
  env2 <- e3[1]
  m1 <- zero2func5a(parvec1, env1)
  m2 <- onestep.4.func(parvec2, env2)
  matconnect(m1, 2, 1, m2)
}

[[2]]$model.obj$comp.name:
[1] "zero2func5"

[[2]]$model.obj$follow.vec:
[1] "0" "0"
```

```
[[2]]$model.obj$workspace.vec:
[1] "zero2func5"    "zero2func5a"    "zero2func00"    "onestep.3.func"
[5] "matconnect"    "onestep.4.func"

[[2]]$fvec:
[1] 0 1 0

[[2]]$model.id:
[1] 5

[[2]]$transient.id:
[1] T

[[2]]$start.vec:
[1] 1 0 0

[[3]]:
[[3]]$model.obj:
[[3]]$model.obj$parameter.vec:
[1] "nu3" "Ea3" "nu4" "Ea4"

[[3]]$model.obj$envir.vec:
[1] "temperature"

[[3]]$model.obj$func:
function(p3, e3)
{
 parvec1 <- p3[c(1:2)]
 parvec2 <- p3[c(c(3, 4))]
 env1 <- e3[1]
 env2 <- e3[1]
 m1 <- zero2func6a(parvec1, env1)
 m2 <- onestep.4.func(parvec2, env2)
 matconnect(m1, 2, 3, m2)
}

[[3]]$model.obj$comp.name:
[1] "zero2func6"

[[3]]$model.obj$follow.vec:
[1] "0" "0"

[[3]]$model.obj$workspace.vec:
[1] "zero2func6"    "zero2func6a"    "zero2func00"    "onestep.3.func"
[5] "matconnect"    "onestep.4.func"

[[3]]$fvec:
[1] 0 0 1

[[3]]$model.id:
[1] 6
```

```
[[3]]$transient.id:
[1] F

[[3]]$start.vec:
[1] 1 0 0


[[4]]:
[[4]]$model.obj:
[[4]]$model.obj$parameter.vec:
[1] "nu3"  "Ea3"  "nu4"  "Ea4"

[[4]]$model.obj$envir.vec:
[1] "temperature"

[[4]]$model.obj$func:
function(p3, e3)
{
 parvec1 <- p3[c(1:2)]
 parvec2 <- p3[c(c(3, 4))]
 env1 <- e3[1]
 env2 <- e3[1]
 m1 <- zero2func7a(parvec1, env1)
 m2 <- onestep.4.func(parvec2, env2)
 matconnect(m1, 2, 3, m2)
}

[[4]]$model.obj$comp.name:
[1] "zero2func7"

[[4]]$model.obj$follow.vec:
[1] "0" "0"

[[4]]$model.obj$workspace.vec:
[1] "zero2func7"   "zero2func7a"   "zero2func00"   "onestep.3.func"
[5] "matconnect"   "onestep.4.func"

[[4]]$fvec:
[1] 0 1 0

[[4]]$model.id:
[1] 7

[[4]]$transient.id:
[1] T

[[4]]$start.vec:
[1] 1 0 0


[[5]]:
[[5]]$model.obj:
[[5]]$model.obj$parameter.vec:
[1] "nu3"  "Ea3"  "nu4"  "Ea4"
```

Evanescent Process Mapping

```
[[5]]$model.obj$envir.vec:
[1] "temperature"

[[5]]$model.obj$func:
function(p3, e3)
{
 parvec1 <- p3[c(1:2)]
 parvec2 <- p3[c(c(3, 4))]
 env1 <- e3[1]
 env2 <- e3[1]
 m1 <- zero2func12a(parvec1, env1)
 m2 <- onestep.4.func(parvec2, env2)
 matconnect(m1, 3, 2, m2)
}

[[5]]$model.obj$comp.name:
[1] "zero2func12"

[[5]]$model.obj$follow.vec:
[1] "0" "0"

[[5]]$model.obj$workspace.vec:
[1] "zero2func12"     "zero2func12a"    "zero2func00"    "onestep.3.func"
[5] "matconnect"      "onestep.4.func"

[[5]]$fvec:
[1] 0 1 0

[[5]]$model.id:
[1] 12

[[5]]$transient.id:
[1] F

[[5]]$start.vec:
[1] 1 0 1
```

For the GUIs pictured in Figure 4.14 and Figure 4.15 the start vectors we chose for the processes were

```
> startvec.list3<-list(c(1,0,0))
> startvec.list4<-list(c(1,0,0,0),c(1,0,0,1),c(1,0,0,1),c(1,0,0,1),
    c(1,0,0,1))
```

The next step in calculation is to create demarcation array calculation functions for each set of models. To access the GUI for this, left-click on "create demarc array calculator for evanescent processes" in the evanescent process drop-down menu. In Figure 4.17 we show it filled in to create the function "zeromod.2.demarc" (this is the string partially shown in the comp.name slot in the GUI). The other three demarcation functions are created similarly, one for each set of functions.

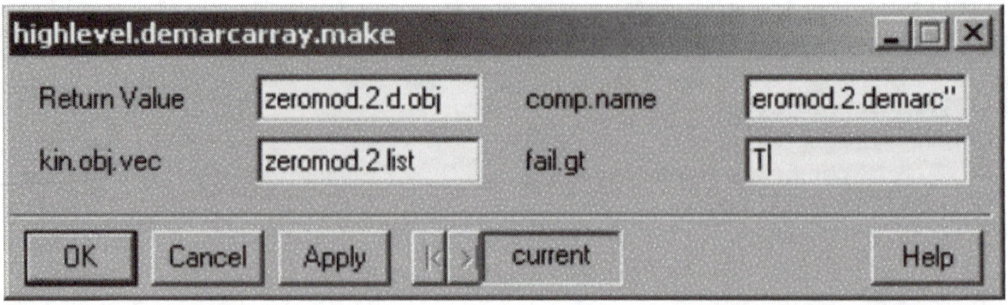

126   *Design and Analysis of Accelerated Tests for Mission Critical Reliability*

```
highlevel.demarcarray.make                              _ □ ×
Return Value    zeromod.2.d.obj      comp.name     eromod.2.demarc"
kin.obj.vec     zeromod.2.list       fail.gt       T|
     OK    Cancel    Apply    |<  >  current                  Help
```

FIGURE 4.17
GUI of the function that writes the demarcation array calculator based on the list of functions selected in one of Figures 4.12, 4.13, 4.15, or 4.16. Here it is shown to operate on the list created in Figure 4.13.

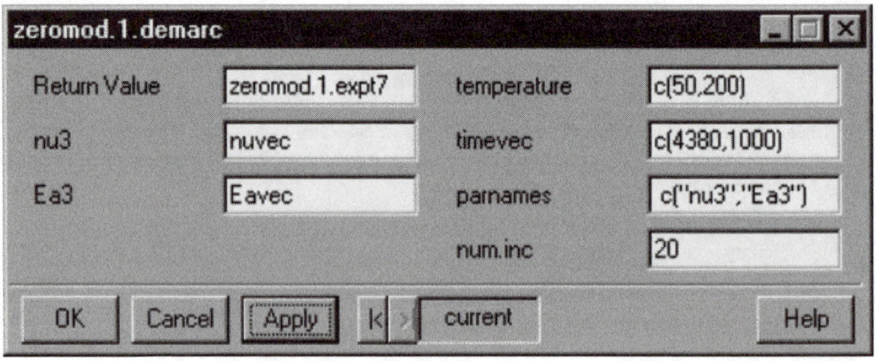

```
zeromod.1.demarc                                        _ □ ×
Return Value    zeromod.1.expt7     temperature    c(50,200)
nu3             nuvec               timevec        c(4380,1000)
Ea3             Eavec               parnames       c("nu3","Ea3")
                                    num.inc        20
     OK    Cancel    Apply    |<  >  current                  Help
```

FIGURE 4.18
GUI of demarcation array calculator for model 1 of Figure 4.4 (follow, Figure 4.12, substitute correctly in Figure 4.17) filled in to calculate demarcation matrix for experiment 7 (Table 4.1).

Each of these functions exists in the functions folder (created in accordance with the instructions in the appendix) after creation, left-clicking on them pops up the GUIs. In Figure 4.18 through Figure 4.21 the four GUIs are filled out to create demarcation map structures for experiment 7.

The parnames vector is a vector of strings, all the parameters listed in the demarcation function. The program structuring results in consistent labeling of the parameters, because the reversible model contains the first two models, with parameters suffixed with 1 and 2; those are the fixed parameters, while the parameters with suffixes 3 and 4 are freely varying.

Once the experiments are created, we concatenate them into structures consistent with the parameter subspaces spanned. The commands are listed below for creating experiment lists, for each of the four affine subspaces.

```
>zeromod.1.exptlist<-list(zeromod.1.expt1,zeromod.1.expt2,
     zeromod.1.expt3,zeromod.1.expt4,zeromod.1.expt5,
     zeromod.1.expt6,zeromod.1.expt7)
```

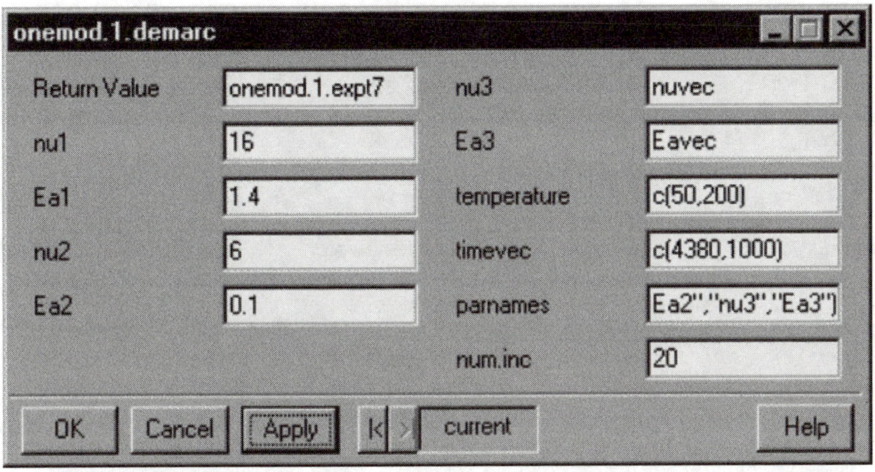

Evanescent Process Mapping

FIGURE 4.19
GUI of demarcation array calculator for model 3 of Figure 4.5 through Figure 4.7 (based on Figure 4.15, substitute correctly in Figure 4.17) filled in to calculate demarcation matrix for experiment 7 (Table 4.1).

FIGURE 4.20
GUI of demarcation array calculator for models 3, 5, 6, 7, and 12 of Figure 4.4 (based on Figure 4.13 and Figure 4.17) filled in to calculate demarcation matrix for experiment 7 (Table 4.1).

```
>zeromod.2.exptlist<-list(zeromod.2.expt1,zeromod.2.expt2,
     zeromod.2.expt3,zeromod.2.expt4,zeromod.2.expt5,
     zeromod.2.expt6,zeromod.2.expt7)
>onemod.1.exptlist<-list(onemod.1.expt1,onemod.1.expt2,
     onemod.1.expt3,onemod.1.expt4,onemod.1.expt5,
     onemod.1.expt6,onemod.1.expt7)
>onemod.2.exptlist<-list(onemod.2.expt1,onemod.2.expt2,
     onemod.2.expt3,onemod.2.expt4,onemod.2.expt5,
     onemod.2.expt6,onemod.2.expt7)
```

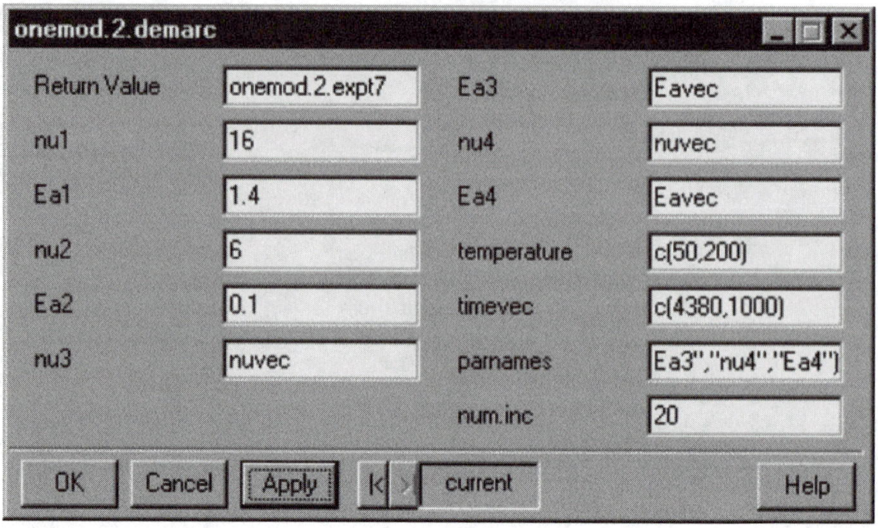

FIGURE 4.21
GUI of demarcation array calculator for models 12, 28, 29, 68, and 70 of Figure 4.5 through Figure 4.7 (based on Figure 4.16, substitute correctly in Figure 4.17) filled in to calculate demarcation matrix for experiment 7 (Table 4.1).

Next we analyze the experiments in terms of the partition constructed in Complement 2.11. To access the GUI left-click on "identify measure for each partition induced by experiments," which brings up the GUI for the function thresh.int.identify. For each of the 2^N partitions for the N experiments described in Complement 2.11, the evaluated parameter values in that partition are integrated with respect to the measure. The function measure.func1 gives a uniform measure. The function measure.funcA works for pure Arrhenius systems and provides the measure where values of the premultiplier below 10^5 Hz are weighted less

$$\exp\left(-\left(\frac{5-\log_{10}(\nu)}{2.5}\right)\right)$$

The program requires the user to identify where the extent values are in the data sets. The simplest way to check this is to print out the first few rows of the "*.life" matrix. Thus, for zeromod.1.*, the first five rows are

```
> zeromod.1.life[1:5,]
      nu3        Ea3       extent1
[1,]  -6   0.0000000   1.000000e+000
[2,]  -6   0.3333333   3.416635e-003
[3,]  -6   0.6666667   1.485718e-008
[4,]  -6   1.0000000   6.449581e-014
[5,]  -6   1.3333333   2.799796e-019
```

Evanescent Process Mapping

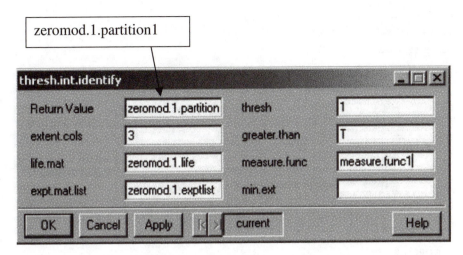

FIGURE 4.22
GUI of function that calculates matrix of posterior measures over the partition of domination of experiments over the life stress trajectory corresponding to Expression 2.23, corresponding a NULL result in the experiment(s) (no failures or noticeable degradation). Here it is filled out to do the calculation for the model 1 from Figure 4.4, using a uniform measure.

The GUI filled out for the uniform measure is shown in Figure 4.22. Then for onemod.2.* the first five rows are

```
> onemod.2.life[1:5,]
     nu1 Ea1 nu2 Ea2 nu3 Ea3 nu4         Ea4      extent1    extent2
[1,]  16 1.4   6 0.1  -6   0  -6 0.0000000 9.262577e-029   2.000000
[2,]  16 1.4   6 0.1  -6   0  -6 0.3333333 9.262577e-029   1.999996
[3,]  16 1.4   6 0.1  -6   0  -6 0.6666667 9.262577e-029   2.000000
[4,]  16 1.4   6 0.1  -6   0  -6 1.0000000 9.262577e-029   2.000000
[5,]  16 1.4   6 0.1  -6   0  -6 1.3333333 9.262577e-029   2.000000
         extent3       extent4       extent5
[1,] 4.052134e-027 1.831098e-011 5.000000e-001
[2,] 2.292796e-016 2.441459e-011 4.341057e-006
[3,] 9.954842e-022 2.441464e-011 4.325937e-011
[4,] 4.414074e-027 2.441464e-011 2.441472e-011
[5,] 9.264453e-029 2.441464e-011 2.441464e-011
```

and the GUI filled out for measure A and onemod.2.partionA is shown in Figure 4.23. The result of applying the second GUI is the structure:

```
> onemod.2.partitionA
$ID.mat:
      [,1] [,2] [,3] [,4] [,5] [,6] [,7]    [,8]         [,9]    [,10]    [,11]        [,12]
 [1,]    1    0    0    0    0    0    0 0.00000 0.000000e+000 0.000000    0.000 0.000000e+000
 [2,]    0    1    0    0    0    0    0 0.00000 1.526535e+001 0.000000    0.000 3.855335e+000
 [3,]    1    1    0    0    0    0    0 0.00000 0.000000e+000 0.000000    0.000 0.000000e+000
 [4,]    0    0    1    0    0    0    0 0.00000 1.008153e+001 0.000000    0.000 0.000000e+000
 [5,]    1    0    1    0    0    0    0 0.00000 0.000000e+000 0.000000    0.000 0.000000e+000
 [6,]    0    1    1    0    0    0    0 0.00000 1.013939e+000 0.000000    0.000 0.000000e+000
 [7,]    1    1    1    0    0    0    0 0.00000 0.000000e+000 0.000000    0.000 0.000000e+000
 [8,]    0    0    0    1    0    0    0 0.00000 1.178651e+001 0.000000    0.000 0.000000e+000
 [9,]    1    0    0    1    0    0    0 0.00000 4.076220e-002 0.000000    0.000 0.000000e+000
[10,]    0    1    0    1    0    0    0 0.00000 6.328447e+000 0.000000    0.000 0.000000e+000
[11,]    1    1    0    1    0    0    0 0.00000 0.000000e+000 0.000000    0.000 0.000000e+000
```

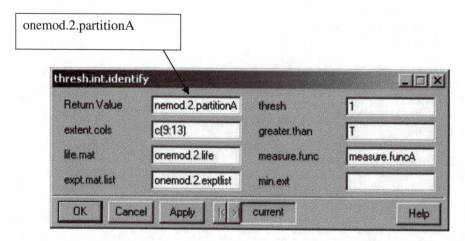

FIGURE 4.23
GUI of same function as in Figure 4.22, Here it is filled out to do the calculation for the models 12, 28, 29, 68, and 70 of Figure 4.5 through Figure 4.7 using a special measure down-weighting low premultiplier values (see step 4 in the list of analysis steps).

```
[12,]  0  0  1  1  0  0  0   0.00000  2.147613e+000   0.000000   0.000  0.000000e+000
[13,]  1  0  1  1  0  0  0   0.00000  0.000000e+000   0.000000   0.000  0.000000e+000
[14,]  0  1  1  1  0  0  0   0.00000  2.000000e+000   0.000000   0.000  0.000000e+000
[15,]  1  1  1  1  0  0  0   0.00000  4.493290e-001   0.000000   0.000  0.000000e+000
[16,]  0  0  0  0  1  0  0   0.00000  1.747138e+001   0.000000   0.000  0.000000e+000
[17,]  1  0  0  0  1  0  0   0.00000  0.000000e+000   0.000000   0.000  0.000000e+000
[18,]  0  1  0  0  1  0  0   0.00000  3.898658e+000   0.000000   0.000  0.000000e+000
[19,]  1  1  0  0  1  0  0   0.00000  0.000000e+000   0.000000   0.000  0.000000e+000
[20,]  0  0  1  0  1  0  0   0.00000  1.227734e-002   0.000000   0.000  0.000000e+000
[21,]  1  0  1  0  1  0  0   0.00000  0.000000e+000   0.000000   0.000  0.000000e+000
[22,]  0  1  1  0  1  0  0   0.00000  0.000000e+000   0.000000   0.000  0.000000e+000
[23,]  1  1  1  0  1  0  0   0.00000  0.000000e+000   0.000000   0.000  0.000000e+000
[24,]  0  0  0  1  1  0  0   0.00000  5.303954e-002   0.000000   0.000  0.000000e+000
[25,]  1  0  0  1  1  0  0   0.00000  2.000000e+000   0.000000   0.000  0.000000e+000
[26,]  0  1  0  1  1  0  0   0.00000  4.861610e-001   0.000000   0.000  0.000000e+000
[27,]  1  1  0  1  1  0  0   0.00000  7.584664e+000   0.000000   0.000  0.000000e+000
[28,]  0  0  1  1  1  0  0   0.00000  1.393890e-002   0.000000   0.000  0.000000e+000
[29,]  1  0  1  1  1  0  0   0.00000  0.000000e+000   0.000000   0.000  0.000000e+000
[30,]  0  1  1  1  1  0  0   0.00000  1.227734e-002   0.000000   0.000  0.000000e+000
[31,]  1  1  1  1  1  0  0   0.00000  4.005517e+000   0.000000   0.000  0.000000e+000
[32,]  0  0  0  0  0  1  0   0.00000  7.585165e+000   0.000000   0.000  2.201897e+000
[33,]  1  0  0  0  0  1  0   0.00000  0.000000e+000   0.000000   0.000  0.000000e+000
[34,]  0  1  0  0  0  1  0   0.00000  4.493290e-001   0.000000   0.000  4.493290e-001
[35,]  1  1  0  0  0  1  0   0.00000  0.000000e+000   0.000000   0.000  0.000000e+000
[36,]  0  0  1  0  0  1  0   0.00000  3.024555e+000   0.000000   0.000  0.000000e+000
[37,]  1  0  1  0  0  1  0   0.00000  1.000000e+000   0.000000   0.000  0.000000e+000
[38,]  0  1  1  0  0  1  0   0.00000  2.584664e+000   0.000000   0.000  0.000000e+000
[39,]  1  1  1  0  0  1  0   0.00000  0.000000e+000   0.000000   0.000  0.000000e+000
[40,]  0  0  0  1  0  1  0   0.00000  1.000000e+000   0.000000   0.000  0.000000e+000
[41,]  1  0  0  1  0  1  0   0.00000  0.000000e+000   0.000000   0.000  0.000000e+000
[42,]  0  1  0  1  0  1  0   0.00000  1.661557e-003   0.000000   0.000  0.000000e+000
[43,]  1  1  0  1  0  1  0   0.00000  0.000000e+000   0.000000   0.000  0.000000e+000
[44,]  0  0  1  1  0  1  0   0.00000  1.140852e+000   0.000000   0.000  0.000000e+000
[45,]  1  0  1  1  0  1  0   0.00000  0.000000e+000   0.000000   0.000  0.000000e+000
[46,]  0  1  1  1  0  1  0   0.00000  2.720000e+000   0.000000   0.000  0.000000e+000
[47,]  1  1  1  1  0  1  0   0.00000  0.000000e+000   0.000000   0.000  0.000000e+000
[48,]  0  0  0  0  1  1  0   0.00000  5.982269e+000   0.000000   0.000  0.000000e+000
[49,]  1  0  0  0  1  1  0   0.00000  4.493290e-001   0.000000   0.000  0.000000e+000
[50,]  0  1  0  0  1  1  0   0.00000  2.675780e+000   0.000000   0.000  0.000000e+000
[51,]  1  1  0  0  1  1  0   0.00000  2.706706e-001   0.000000   0.000  0.000000e+000
[52,]  0  0  1  0  1  1  0   0.00000  4.401014e-002   0.000000   0.000  0.000000e+000
[53,]  1  0  1  0  1  1  0   0.00000  0.000000e+000   0.000000   0.000  0.000000e+000
[54,]  0  1  1  0  1  1  0   0.00000  2.024555e+000   0.000000   0.000  0.000000e+000
[55,]  1  1  1  0  1  1  0   0.00000  1.418283e+000   0.000000   0.000  0.000000e+000
[56,]  0  0  0  1  1  1  0   0.00000  4.565896e+000   0.000000   0.000  0.000000e+000
[57,]  1  0  0  1  1  1  0   0.00000  0.000000e+000   0.000000   0.000  0.000000e+000
[58,]  0  1  0  1  1  1  0   0.00000  4.181391e+000   0.000000   0.000  0.000000e+000
[59,]  1  1  0  1  1  1  0   0.00000  7.936859e+000   0.000000   0.000  0.000000e+000
[60,]  0  0  1  1  1  1  0   0.00000  1.065317e+000   0.000000   0.000  0.000000e+000
```

Evanescent Process Mapping

```
 [61,]   1  0  1  1  1  1  0     0.00000  2.947347e+000     0.000000     0.000  0.000000e+000
 [62,]   0  1  1  1  1  1  0     0.00000  6.018941e+000     0.000000     0.000  0.000000e+000
 [63,]   1  1  1  1  1  1  0     0.00000  3.598792e-001     0.000000     0.000  0.000000e+000
 [64,]   0  0  0  0  0  0  1     0.00000  9.058548e+000     0.000000     0.000  0.000000e+000
 [65,]   1  0  0  0  0  0  1     0.00000  4.616063e-001     0.000000     0.000  0.000000e+000
 [66,]   0  1  0  0  0  0  1     0.00000  3.000000e+000     0.000000     0.000  0.000000e+000
 [67,]   1  1  0  0  0  0  1     0.00000  1.227734e-002     0.000000     0.000  0.000000e+000
 [68,]   0  0  1  0  0  0  1     0.00000  7.449329e+000     0.000000     0.000  0.000000e+000
 [69,]   1  0  1  0  0  0  1     0.00000  0.000000e+000     0.000000     0.000  0.000000e+000
 [70,]   0  1  1  0  0  0  1     0.00000  0.000000e+000     0.000000     0.000  0.000000e+000
 [71,]   1  1  1  0  0  0  1     0.00000  1.353353e-001     0.000000     0.000  0.000000e+000
 [72,]   0  0  0  1  0  0  1     0.00000  7.390271e+000     0.000000     0.000  0.000000e+000
 [73,]   1  0  0  1  0  0  1     0.00000  1.760975e-001     0.000000     0.000  0.000000e+000
 [74,]   0  1  0  1  0  0  1     0.00000  2.449329e+000     0.000000     0.000  0.000000e+000
 [75,]   1  1  0  1  0  0  1     0.00000  1.000000e+000     0.000000     0.000  0.000000e+000
 [76,]   0  0  1  1  0  0  1     0.00000  4.625426e+000     0.000000     0.000  0.000000e+000
 [77,]   1  0  1  1  0  0  1     0.00000  1.000000e+000     0.000000     0.000  0.000000e+000
 [78,]   0  1  1  1  0  0  1     0.00000  5.000000e+000     0.000000     0.000  0.000000e+000
 [79,]   1  1  1  1  0  0  1     0.00000  1.961453e-001     0.000000     0.000  0.000000e+000
 [80,]   0  0  0  0  1  0  1     0.00000  0.000000e+000     0.000000     0.000  0.000000e+000
 [81,]   1  0  0  0  1  0  1     0.00000  7.760762e+000     0.000000     0.000  0.000000e+000
 [82,]   0  1  0  0  1  0  1     0.00000  1.000000e+000     0.000000     0.000  0.000000e+000
 [83,]   1  1  0  0  1  0  1     0.00000  1.000000e+000     0.000000     0.000  0.000000e+000
 [84,]   0  0  1  0  1  0  1     0.00000  1.449329e+000     0.000000     0.000  0.000000e+000
 [85,]   1  0  1  0  1  0  1     0.00000  1.353353e-001     0.000000     0.000  0.000000e+000
 [86,]   0  1  1  0  1  0  1     0.00000  1.135335e+000     0.000000     0.000  0.000000e+000
 [87,]   1  1  1  0  1  0  1     0.00000  1.176097e+000     0.000000     0.000  0.000000e+000
 [88,]   0  0  0  1  1  0  1     0.00000  1.135335e+000     0.000000     0.000  0.000000e+000
 [89,]   1  0  0  1  1  0  1     0.00000  1.135335e+000     0.000000     0.000  0.000000e+000
 [90,]   0  1  0  1  1  0  1     0.00000  1.227734e-002     0.000000     0.000  0.000000e+000
 [91,]   1  1  0  1  1  0  1     0.00000  1.811552e+001     0.000000     0.000  0.000000e+000
 [92,]   0  0  1  1  1  0  1     0.00000  2.012277e+000     0.000000     0.000  0.000000e+000
 [93,]   1  0  1  1  1  0  1     0.00000  1.000000e+000     0.000000     0.000  0.000000e+000
 [94,]   0  1  1  1  1  0  1     0.00000  0.000000e+000     0.000000     0.000  0.000000e+000
 [95,]   1  1  1  1  1  0  1     0.00000  9.030593e+000     0.000000     0.000  0.000000e+000
 [96,]   0  0  0  0  0  1  1     0.00000  3.593230e+001     0.000000     0.000  6.924143e+001
 [97,]   1  0  0  0  0  1  1     0.00000  1.098636e+000     0.000000     0.000  0.000000e+000
 [98,]   0  1  0  0  0  1  1     0.00000  5.516564e-003     0.000000     0.000  0.000000e+000
 [99,]   1  1  0  0  0  1  1     0.00000  0.000000e+000     0.000000     0.000  0.000000e+000
[100,]   0  0  1  0  0  1  1   953.34868  2.980120e+002   280.470927     0.000  4.275408e+000
[101,]   1  0  1  0  0  1  1     0.00000  9.000000e+000     0.000000     0.000  0.000000e+000
[102,]   0  1  1  0  0  1  1   230.51773  7.596942e+000    86.597230     0.000  1.370765e+000
[103,]   1  1  1  0  0  1  1     0.00000  4.000000e+000     0.000000     0.000  0.000000e+000
[104,]   0  0  0  1  0  1  1     0.00000  5.081524e+000     0.000000     0.000  0.000000e+000
[105,]   1  0  0  1  0  1  1     0.00000  0.000000e+000     0.000000     0.000  0.000000e+000
[106,]   0  1  0  1  0  1  1     0.00000  4.147613e+000     0.000000     0.000  0.000000e+000
[107,]   1  1  0  1  0  1  1     0.00000  4.449329e+000     0.000000     0.000  0.000000e+000
[108,]   0  0  1  1  0  1  1     0.00000  5.429128e+001     2.515244     0.000  0.000000e+000
[109,]   1  0  1  1  0  1  1     0.00000  1.510139e+000     0.000000     0.000  0.000000e+000
[110,]   0  1  1  1  0  1  1   195.97236  5.580371e+001   253.069727     0.000  0.000000e+000
[111,]   1  1  1  1  0  1  1     0.00000  2.708026e+001     0.000000     0.000  0.000000e+000
[112,]   0  0  0  0  1  1  1     0.00000  1.596942e+000     0.000000     0.000  0.000000e+000
[113,]   1  0  0  0  1  1  1     0.00000  1.339378e+001     0.000000     0.000  4.493290e-001
[114,]   0  1  0  0  1  1  1     0.00000  1.227734e-002     0.000000     0.000  0.000000e+000
[115,]   1  1  0  0  1  1  1     0.00000  4.065244e+001     0.000000     0.000  1.510269e+001
[116,]   0  0  1  0  1  1  1     0.00000  3.683202e-002     0.000000     0.000  0.000000e+000
[117,]   1  0  1  0  1  1  1     0.00000  1.625469e+001     0.000000     0.000  0.000000e+000
[118,]   0  1  1  0  1  1  1     0.00000  4.270671e+000     0.000000     0.000  1.326506e+001
[119,]   1  1  1  0  1  1  1     0.00000  1.334744e+001     0.000000     0.000  1.353353e-001
[120,]   0  0  0  1  1  1  1     0.00000  1.648665e+000     0.000000     0.000  0.000000e+000
[121,]   1  0  0  1  1  1  1     0.00000  3.833481e+001     0.000000     0.000  7.535866e+001
[122,]   0  1  0  1  1  1  1     0.00000  5.388749e+000     0.000000     0.000  0.000000e+000
[123,]   1  1  0  1  1  1  1     0.00000  7.639376e+002     0.000000     0.000  4.997960e+002
[124,]   0  0  1  1  1  1  1     0.00000  2.465230e+001     1.622866     0.000  0.000000e+000
[125,]   1  0  1  1  1  1  1     0.00000  5.959523e+001    24.362440     0.000  0.000000e+000
[126,]   0  1  1  1  1  1  1    77.78355  2.721298e+001    75.662199     0.000  1.507331e-004
[127,]   1  1  1  1  1  1  1  2948.28885  2.531412e+003  3681.610530  4405.911  3.645899e+003
[128,]   0  0  0  0  0  0  0     0.00000  8.837585e+001     0.000000     0.000  8.764354e+001

$tot:
   extent1   extent2   extent3   extent4   extent5
  4405.911  4405.911  4405.911  4405.911  4405.911

$ncols:
[1] 5

$nlist:
[1] 7
```

The partitions are listed by which experiments dominate life in the partition in ID.mat. The first seven columns represent experiments 1 through 7; a 1 in the column indicates the experiment dominates life. Thus, line 127

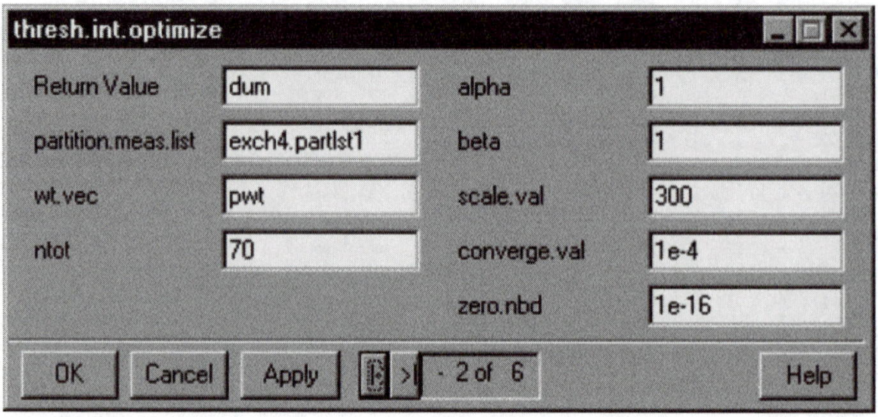

FIGURE 4.24
GUI to "optimize" distribution of sample weights (sample sizes) across experiments. Note that scale.val, converge.val, and zero.nbd are parameters tuning parameters for the optimization algorithm and may require some fiddling.

shows the measure for each model associated with the partition where every one of the experiments dominates life.

The component of the list labeled tot shows the total measure when you sum the partition, so the value of $\pi(P_k)$ from Complement 2.11 for the model corresponding to extent 1 for that partition is 2948.28885/4405.911 = 0.66916666.

For each measure we now combine the calculated partitions as below:

```
> exch4.partlst1<-list(zeromod.1.partition1,onemod.1.partition1,
      zeromod.2.partition1,onemod.2.partition1)
>exch4.partlstA<-list(zeromod.1.partitionA,onemod.1.partitionA,
      zeromod.2.partitionA,onemod.2.partitionA)
```

Finally, we can calculate the weights we wish to use for each experiment to optimize them. Thus, to do the calculation (based on Complement 2.11) to fill in the first row of Table 4.2, we pull up the GUI for the function "thresh.int.optimize" by left-clicking on the heading "optimize nvec to minimize posterior probability for NULL experiment" under the "evanescent process maps" drop-down menu, and fill it out as in Figure 4.24.

To create a full specification of a prior, we need to provide a weight for the models. Assuming that simpler models are more likely, model 1 from Figure 4.4 and (corresponding to zeromod.1*) and model 3 from Figure 4.5 (corresponding to onemod.1.* in the above) require only one extra arrow, so we can put half of the probability split evenly between the two. The other ten models all require two more arrows, so we provide split the other half of the probability between them. Based on the order in the list, this results in a prior weight vector of the form:

Evanescent Process Mapping

```
pwt
 [1]  0.25  0.25  0.05  0.05  0.05  0.05  0.05  0.05  0.05  0.05  0.05  0.05
```

The values of scale.val, converge.val, and zero.nbd determine convergence and convergence rate in the algorithm. As the program operates, it prints a list of the value being compared to converge.val. Scale.val controls the rate of the search; default is 100. The program as set up stops when the posterior probabilities show a relative change of less than 1e–4.

The output of the program above is shown below:

```
dum
$wt.vec:
 [1]  0.003091011  0.133473082  0.088010939  0.043251049  0.048909315
 [6]  0.285447731  0.397816873

$post.prob:
 [1]  0.03192625

$post.prob0:
 [1]  0.03271157

$irreducible:
 [1]  0.01729703

$Ntot:
 [1]  70
```

The irreducible portion is the prior probability corresponding to the partition where life dominates the experiment. We point out that the total measure corresponding to the irreducible portion is twice this, but the uncertainty added by the prior β distribution reduces that by a factor of two.

5

Data Analysis for Failure Time Data

To this point we have focused on the design and interpretation of experiments in which no measurable degradation or failures have occurred. For systems where components are mission-critical single points of failure (failure of that component causes failure of the mission), NULL data of this sort are the best kind of data. In many cases, however, failures or degradation are observed even in life and are part of the design cost trade-off in building a system. It is necessary in those cases to be able to analyze the data. In accelerated testing problem there are several kinds of data:

1. Failure time data: For each experimental unit the observable is the time at which it fails.
2. Degradation data: For each experimental unit the observable is a set of measurements over time, and there is a gradual change in that set of measurements.
3. "Breaking strength" data: At given times a sample of experimental unit from a larger sample exposed to stress is subject to a destructive set of measurements.

There are a number of texts discussing the analysis of failure time data (Nelson, 1982; Lawless, 1982; Meeker and Escobar, 1998; among others) and some contain discussion of degradation data. However, none uses kinetic models as the starting point of the discussion. In this and the following chapter we discuss some ways to use kinetic models to analyze degradation and failure data. With failure time data, the only data on a device are the device pedigree and time to fail or end of test. With degradation data, typically the first job is to extract a meaningful reduction of the very large amount of data to focus on the degradation process. Breaking strength data falls between these, but lies beyond the scope of further discussion in this book.

In reliability texts dealing with accelerated testing, the emphasis is on describing and developing theory and techniques for the analysis of failure time data. In this text the approach is to treat failure time data as one of many possible windows to the degradation processes that ultimately lead to failure.

This chapter introduces the reader to the statistical analysis of failure time data using physical models to complement statistical methods. We begin with an example that can be worked by hand. The analysis is purposely tied to physical reasoning about the problem at hand. We examine data where simple failure time models hold, beginning with some artificial data similar to a real problem that has been encountered. We analyze the data using simple statistical methods, and a simple stochastic model first, then develop connections between that stochastic model and chemical kinetics. From this, a general approach to the analysis of accelerated failure time data, connecting kinetics with statistical behavior, is developed. Finally, a real data set (see Complements 5.8 and 5.9) is analyzed using a chemical kinetic model. The stochastic model has the advantage of being more statistically efficient for small samples, but is unfortunately less flexible than the full kinetic framework.

5.1 A Simple Data Set

Consider a sequence of accelerated tests run at three temperatures, 120°C, 85°C, and 65°C, for 500 h. Some devices fail and others survive the duration of the test, giving censored data. The data are presented below in a structure typically used in Splus. There are three data sets, each comprising three columns. The first column is either time to failure in hours or the duration of the experiment (500 h) if the device does not fail. The second is an indicator, 1 for failure or 0 for survival. The third column is the temperature of the experiment in degrees centigrade. The double brackets denote a component of a list in Splus.

```
[[1]]
                v1
 [1,]  0.08174556  1  120
 [2,]  0.10466682  1  120
 [3,]  0.25828953  1  120
 [4,]  0.30636140  1  120
 [5,]  0.31807719  1  120
 [6,]  0.41352648  1  120
 [7,]  0.41749921  1  120
 [8,]  0.64156169  1  120
 [9,]  0.64505005  1  120
[10,]  1.09147921  1  120
[11,]  1.43227676  1  120
[12,]  1.48200924  1  120
[13,]  1.57265149  1  120
[14,]  2.15669128  1  120
[15,]  2.38259925  1  120
```

Data Analysis for Failure Time Data

```
[[2]]:
              v1
 [1,]  500.0000000 0 85
 [2,]  500.0000000 0 85
 [3,]  500.0000000 0 85
 [4,]  500.0000000 0 85
 [5,]    5.1278233 1 85
 [6,]    6.9495705 1 85
 [7,]   11.7801221 1 85
 [8,]   35.6629020 1 85
 [9,]   13.3299067 1 85
[10,]   32.1068739 1 85
[11,]   81.7631735 1 85
[12,]   32.3651371 1 85
[13,]   22.4859659 1 85
[14,]    0.4898405 1 85
[15,]   45.3504793 1 85


[[3]]:
              v1
 [1,]   13.00293 1 65
 [2,]   25.71679 1 65
 [3,]   55.56006 1 65
 [4,]  348.16794 1 65
 [5,]  500.00000 0 65
 [6,]  500.00000 0 65
 [7,]  500.00000 0 65
 [8,]  500.00000 0 65
 [9,]  500.00000 0 65
[10,]  500.00000 0 65
[11,]  500.00000 0 65
[12,]  500.00000 0 65
[13,]  500.00000 0 65
[14,]  500.00000 0 65
[15,]  500.00000 0 65
```

A common and often physically justifiable lifetime distribution used in the analysis of failure time data is the Weibull distribution. This distribution has physical meaning as the limiting form of the distribution of time to first failure of a number of independent devices from groups with identical failure distributions. The cumulative life distribution function, the probability that a failure occurs by time t in a device, for the Weibull has the form:

$$P(T \le t) = F(t) = 1 - \exp\left[-\left(\frac{t}{\alpha}\right)^{\beta}\right] \quad (5.1)$$

where T is the failure time of a device (a random variable) and $P(T \le t)$ is the probability that failure occurs in less than time t.

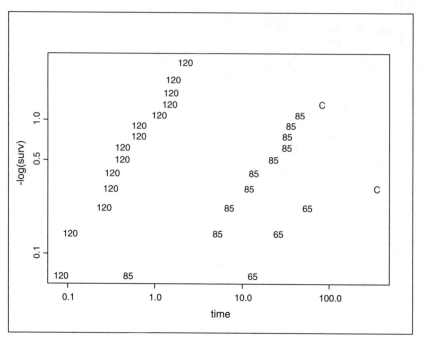

FIGURE 5.1
Simultaneous Weibull plot of data from first example.

The parameters α and β are, respectively, the scale and shape parameters of the distribution. A plot of $\log(-\log(1-F(t)))$ vs. $\log(t)$ (Figure 5.1, note $1 - F(t)$ is the survival function, hence "surv" in the plot) gives a straight line with slope β if F has the form given in Equation 5.1. This type of plot has an interesting property noted by Lawless (1982), which requires some background to understand. Note that the points labeled C represent the last failed devices before censoring begins in Figure 5.1.

The failure time distribution depends on the accelerating "stress," in this case temperature. There are two "standard" models for explaining the dependence of failure time distributions on stress. One is the *accelerated failure time model* where distributions at two conditions are related by a time scaling factor:

$$F_2(t) = F_1(a_{21}t) \qquad (5.2)$$

where $F_1(t)$ is the distribution at condition 1, $F_2(t)$ is the distribution at condition 2, and α_{21} is the time scaling factor.

The other is the *proportional hazards model*, in which the hazard rate, the instantaneous failure probability of a device, given that it has not failed before that time, is scaled with the environment. If we define the lifetime probability density function as $f(t) = (d/dt)F$ then the hazard rate is

Data Analysis for Failure Time Data

$$h(t) = \frac{f(t)}{1-F(t)}$$

The proportional hazards model then has the form:

$$h_2(t) = a_{12} h_1(t) \tag{5.3}$$

The *accelerated failure time model* can be shown to correspond physically to situations where chemistry of the failure mode is dominated by a single rate-limiting step. The *proportional hazards model*, on the other hand, corresponds to conditions where the device weakens over time, but the rate-limiting part of failure is due to the random occurrence of an event of sufficient energy to initiate a relatively fast failure process. The change in environment then corresponds to a change in the nature of the random accidents rather than a change in the weakening rate of the device. This model seems more appropriate to the study of field data, or of biological processes, than to the study of accelerated tests.

The interesting insight of Lawless was that when the life distributions are plotted on a Weibull scale, the proportional hazards model will result in failure distribution curves that are parallel but shift along the vertical axis, while the accelerated life model will result in failure curves that are parallel but shifted along the horizontal axis. The shifts are independent of how well the distribution is fit by a Weibull. Thus, in addition to providing a visual clue whether a failure distribution is Weibull or not (linear on the plot), the plot provides clues regarding what form of acceleration model is appropriate.

Mathematical Readers' Exercise

> Using the definitions of the Weibull plot, accelerated failure time model, and proportional hazards model, demonstrate Lawless's observation.

Our plot (Figure 5.1) shows that the Weibull distribution appears to be a reasonable approximation to the highest stress condition (120°C), but the two other data sets cannot be related to either the accelerated failure time model or the proportional hazards model because the change is not a simple shift (the circled points in the plot represent the last failure before censoring at 500 h of all the remaining devices). The plots suggest that failure times are being stretched more for longer times than for shorter times as stress drops. The simplest way to create this is to stretch the longest times to infinity. A simplistic stochastic model for this is as follows:

Let X be the activation time for the mechanism causing failure (a random variable), Y the corresponding time for the censoring mechanism that does not cause failure (also a random variable), and C the time at which the test is terminated. If $X < Y$ and $X < C$, then we see a failure. Otherwise we do not. We call this the unobservable censoring (UC) model. We can express this in terms of the Weibull parameters after first noting that at the highest stress condition, the slope of the line, i.e., β, is very close to 1. In this special case, the Weibull becomes an exponential distribution of the form:

$$P(T < t) = 1 - \exp(-\lambda t)$$

where $\lambda = 1/\alpha$ from the Weibull model with $\beta = 1$.
For the UC model we can now write

$$P(X < x) = 1 - \exp(-\lambda_1 x)$$

$$P(Y < y) = 1 - \exp(-\lambda_2 y)$$

where $P(X < x)$ is the probability that the failure mechanism occurs in a time less than x and $P(Y < y)$ the probability that the censoring mechanism occurs in a time less than y.

The probability of seeing a failure at a particular time y (for $y \leq C$) with no censoring by the competing mechanism is then

$$P(X < Y) = \int_0^y f_x(x)(1 - F_y(x))dx = \int_0^y \lambda_1 \exp(-\lambda_1 x)(\exp(-\lambda_2 x))dx$$

$$= \int_0^y \lambda_1 \exp(-(\lambda_1 + \lambda_2)x)dx = \frac{\lambda_1}{\lambda_1 + \lambda_2}(1 - \exp(-(\lambda_1 + \lambda_2)y)) \quad (5.4)$$

We now have two potential models: one based on a single mechanism, which generates failure, the other based on two mechanisms, one of which competes and prevents failure. Both are assumed to have exponential life distributions. To test which model is most reasonable, we call on the methodology of maximum likelihood. Complement 5.4 provides a nonrigorous development of the theory of maximum likelihood. Here we assume that theory.

We start by determining the maximum likelihood estimates, and the likelihood for each. For the simple exponential model the likelihood at each condition has the form:

$$l(\vec{t}, \lambda) = \prod_{i \in M}(\lambda \exp(-\lambda t_i)) \prod_{i \in C} \exp(-\lambda t_i) = \lambda^{o(M)} \exp\left(-\lambda \sum_{i=1}^n t_i\right) \quad (5.5)$$

Here we have each t_i representing the failure or censoring time of each device, M is the set of devices failing (M = mortality), C is the set of devices that are censored (recall that $1 - F(t_i)$ is the probability of survival beyond

Data Analysis for Failure Time Data

time t_i), $o(M)$ is the number of devices in set M, and n is the total number of devices. From this we find the log likelihood:

$$L(\vec{t}, \lambda) = o(M)\ln(\lambda) - \lambda \sum_{i=1}^{n} t_i \tag{5.6}$$

To find the maximum value of this with respect to λ, we take the derivative, set it equal to 0, and solve for λ. This gives us

$$\hat{\lambda} = \frac{o(M)}{\sum_{i=1}^{n} x_i} \tag{5.7}$$

To check that this is in fact a maximum, just substitute Equation 5.7 into Equation 5.6, and check the neighborhood.

For the UC model:

$$l(\vec{x}, (\lambda_1, \lambda_2)) = \prod_{i \in M} \left(\lambda_1 \exp\left(-(\lambda_1 + \lambda_2)x_i\right) \right) \\ \times \prod_{i \in C} \left(\frac{\lambda_1}{\lambda_1 + \lambda_2} \exp\left(-(\lambda_1 + \lambda_2)x_i\right) + \frac{\lambda_2}{\lambda_1 + \lambda_2} \right) \tag{5.8}$$

Reducing this and taking the log we obtain

$$L(\vec{t}, (\lambda_1, \lambda_2)) = o(M)\ln(\lambda_1) - (\lambda_1 + \lambda_2)\sum_{i \in M} t_i \\ + \sum_{i \in C} \log\left(\frac{\lambda_1}{\lambda_1 + \lambda_x} \exp\left(-(\lambda_1 + \lambda_2)t_i\right) + \frac{\lambda_2}{\lambda_1 + \lambda_2} \right) \tag{5.9}$$

This does not reduce to simple closed-form expressions; however, it is a simple enough task to code the above formula into a statistical language (such as Splus or SAS®) or a spreadsheet and search for an optimum.

Table 5.1 is an Excel spread sheet, constructed to find maximum likelihood estimates for the exponential and UC models, and to do a statistical test for which model is best.

The data here are from 85°C. The first 11 numbers in the first column are the failure times. The 12th number is the sum of the failure times. The first 11 numbers in the second column are simply indicators of failure, and the 12th is the sum of the indicators. The first four numbers in the fourth column are the censoring times, with the fifth number the sum of those times. The number labeled expml is simply Equation 5.7 calculated using these sums,

TABLE 5.1

Spread Sheet for Maximum Likelihood Calculations

0.49	1		500	0.25641	-1.36098		
5.12	1		500	0.25641	-1.36098		
6.94	1		500	0.25641	-1.36098		
11.78	1		500	0.25641	-1.36098		
35.66	1		2000		-5.44391		
13.32	1						
32.1	1						
81.76	1						
32.36	1						
22.48	1						
45.37	1						
287.38	11						
expml	0.004809						
loglikexp	-69.7099						
unobscl1	0.029					int1	0.74359
unobscl2	0.01					int2	0.039
						int3	0.25641
loglikUC	-55.5968						
chisqtest	28.22632						
Pval	7.43E-07						

i.e., the maximum likelihood estimator of λ for the exponential model. The number labeled loglikexp is the maximum value of log likelihood obtained by substituting the maximum likelihood estimator of λ back into Equation 5.6. The numbers labeled unobscl1 and unobscl2 are trial values for λ_1 and λ_2, respectively. The numbers labeled int1, int2, and int3 are intermediate value calculations: $\lambda_1/(\lambda_1+\lambda_2)$, $\lambda_1+\lambda_2$, and $\lambda_2/(\lambda_1+\lambda_2)$, respectively. The four numbers in the fifth column are the calculation of the term in parentheses on the far right of Equation 5.8 for each censoring value, the first four numbers in the sixth column are their natural logs, and the fifth number in the sixth column is the sum of the logs. The number labeled loglikUC is simply Equation 5.9, the corresponding value of log likelihood for the UC model calculated using these terms. The likelihood for the UC model can be maximized using a solver built in Excel; however, simple trial and error was used here to obtain $\hat{\lambda}$, our estimate of λ.

We now have values of loglikexp (–69.709) and loglikUC (–55.5968) with which to compare the two models. Because the UC model gives the larger value, it appears to be the better model. However, because the log likelihood values are themselves random variables, we need to apply a statistical test to determine the probability that the observed differences in value are real. The statistic we will use is the number labeled chisqtest = $-2\left(L(\bar{x},\lambda) - L(\bar{x},(\lambda_1,\lambda_2))\right)$. According to the theory described in the appendix, we can conservatively assumed that this has a chi-square distribution with two degrees of freedom. (*Note:* There are technical reasons why it is inappropriate to use the chi square distribution with one degree of freedom, which the reader can find reference to in the complement and further study in the references.) The value of chisqtest for this experiment is 28.22632. The probability of a value this large occurring by chance (Pval) is on the order of 1e–6. In other words, the evidence is strong that the UC model is the better of the two for the 85°C data. (Note, however, that other models could be considered and may prove an even better fit to the data.)

Interestingly, the 85°C data comprise the only single data set here where the evidence is strong enough to convince us that the exponential is wrong relative to the UC model. At 65°C, the estimates for λ_1 and for λ_2 are, respectively, 0.0023 and 0.0062, with a log likelihood of –31.47, as opposed to –33.21 for the exponential model. Although the value for the UC model is again greater than that of the exponential, the values of chisqtest and Pval are 3.48 and 0.18. This is not strong evidence in favor of the UC model. At 120°C, all the units fail, so the Y values are all larger than the X values, and the data contain no information concerning the censoring process if not combined with data from other conditions.

At this point, in a real experiment, it would be worthwhile to recheck all the experimental records relating to the 85°C data set for possible anomalies, as it is the single experiment supporting the UC model. Assuming this has been done and no problems are found, a next logical step is to see if we can modify the exponential or the UC models by parametrically connecting all three temperature conditions to obtain a better fit to the data. This is done as follows.

If we assume that our λ_i represent two statistically independent chemical processes that interact through the censoring mechanism, we start by looking for their dependence on temperature. The simplest assumption is that each λ_i has an Arrhenius temperature dependence, $\lambda_i = v\exp(-E_a/kT)$. Plotting the log of λ_1, derived from the UC model, against $1/kT$ (Figure 5.2) we see a straight line.

From this we extract E_a = 1.29 eV, nu(v) = 3.76e16. Doing the same for λ values derived from the exponential model (Figure 5.3) gives E_a = 1.57, v = 1.3e20. The fit for the exponential model is not quite as good, and the value of v is high. This is an artificial problem, but the authors have never seen a value of v quoted that was that high (e.g., Benson, 1976).

Although we have only two good estimates of λ_2 from the experiments (recall that all units failed at 120°C), we can estimate the Arrhenius parameters in the same manner. The result gives an activation energy of 0.25 eV and a value of v of 3e1 Hz.

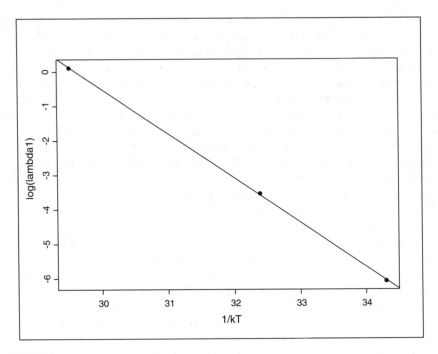

FIGURE 5.2
Plot of log(lambda1) from the UC model vs. $1/kT$.

The above estimates for activation energy and λ are based on linear fits to the Arrhenius relationship of λ values, which are maximum likelihood estimates treating the data sets from each temperature independently. We can calculate a *joint* likelihood for all three data sets, by assuming the Arrhenius relationship to describe $\lambda_1, \lambda_2,$ and λ.

For the exponential model this gives us two parameters (a single activation energy and a v value) to fit all three data sets, whereas for the UC model there are four parameters to fit all three data sets (an activation energy and a v value for each λ_i). It is possible to set up an Excel sheet for the full likelihood calculation; however, the estimates and likelihoods below were calculated using functions specially written in Splus for this purpose. The source code for these Splus programs are provided in Complement 5.4. The optimization was done using the optimization code in our freeware, which is available in Splus.

The optimized parameter values are shown below:

Exponential model:

$$\log_{10}(v) = 22.23, \; E_a = 1.74, \; L(\bar{x}, \theta) = -118.4$$

UC model:

Data Analysis for Failure Time Data

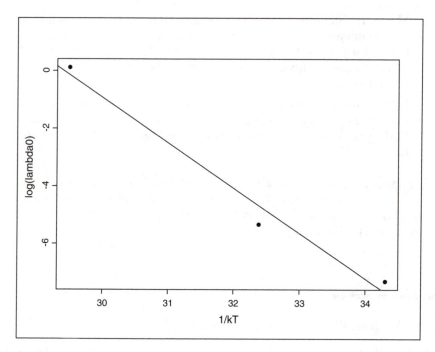

FIGURE 5.3
Plot of log(lambda0) from the exponential model vs. $1/kT$.

$$\left.\begin{array}{l}\log_{10}(v_1) = 16.54, \ E_{a1} = 1.29 \\ \log_{10}(v_2) = 1.14, \ E_{a2} = 0.224\end{array}\right\} L(\vec{x}, \theta) = -100.51$$

The likelihood is greater for the UC model, but before accepting it we again need a statistical test. Using the chisqtest statistic, as before, we find that its value (two times the difference in log likelihoods) is >35. The corresponding Pval for the UC model, assuming the exponential model is true, is the tail probability of a chi-square distribution, with four degrees of freedom, which in this case is <1e–6. It is highly improbable that this would have happened by chance if the exponential model were in fact true; therefore, we reject the exponential model.

Before examining the physical and kinetics aspects of the UC model, we summarize here what we have done to this point:

- An initial examination of the data on Weibull plots led to the postulation of two potential models: the exponential and the UC, each with its own form and parameters in the life distribution function, $F(t)$.
- $F(t)$ for each model was expressed as a function of temperature by assuming an Arrhenius relation

- Maximum likelihood analysis was used to derive best estimates for the parameters and the corresponding maximum value of the likelihood for each of the models. On the basis of the greater value of its maximum likelihood, the UC model was accepted as the better of the two.
- Because the parameter estimates for the UC were derived as a function of temperature, instantaneous failure rate (FITs) estimates can be derived at any service temperature and time by use of the hazard function $h(t)$.

The accuracy of the failure rate estimates is determined by the validity of the model and the number of data points. The latter affects the confidence interval of the estimate, a subject that is treated later (see Section 5.3 and Complements 5.5 and 5.6).

5.2 Adding Physical Sense to the Model

The model provided above is mathematically consistent and fits the data, but it is purely empirical. It would be more complete and have greater validity if it were also based on principles of physical chemistry. For example, it tells us nothing about what would happen if in the middle of the experiment we suddenly raised the temperature, because we have not tied it to a physical understanding of the process.

An alternative to the approach taken in Section 5.1 is to begin with chemical models as a basis for statistical modeling (LuValle et al., 1988). It happens that there are two schema for chemical models that give the same distribution as Equation 5.4. The chemical models specified by these schema belong to the class of linear first-order chemical kinetic processes that we have used as a basis for working with demarcation and evanescent process maps.

If a uniform distribution of thresholds leading to failure is assumed, and no initial material in the failure-causing state, designated A_f, then the failure distributions given in Equation 5.4 can be derived from the following two different kinetic models. The differential equations and weighted directed graphs for each system of equations are given below.

Model 1: Equilibration

$$\frac{d}{dt}\begin{pmatrix} A_{1t} \\ A_{2t} \end{pmatrix} = \begin{pmatrix} -k_1 & k_2 \\ k_1 & -k_2 \end{pmatrix}\begin{pmatrix} A_{1t} \\ A_{2t} \end{pmatrix} \qquad A_1 \underset{k_2}{\overset{k_1}{\rightleftarrows}} A_2$$

Model 2: Competing reactions

$$\frac{d}{dt}\begin{pmatrix} A_{1t} \\ A_{2t} \\ A_{3t} \end{pmatrix} = \begin{pmatrix} -(k_1+k_2) & 0 & 0 \\ k_1 & 0 & 0 \\ k_2 & 0 & 0 \end{pmatrix} \begin{pmatrix} A_{1t} \\ A_{2t} \\ A_{3t} \end{pmatrix} \qquad A_1 \begin{array}{c} \xrightarrow{k_1} A_2 \\ \xrightarrow{k_2} A_3 \end{array}$$

In both cases we assume that if A_2 exceeds a threshold (uniformly distributed between devices) the device under test fails. Also in both cases, simple *constant* temperature accelerated tests will result in failure distributions of the kind we model above. However, step stress experiments, in which the device is stressed at one temperature, then moved to another temperature in the middle of the experiment, give quite different results (LuValle and Hines, 1992). In particular, suppose in both models that k_2 has a lower activation energy than k_1. If they have different premultipliers associated with them, the probability of failure will drop with temperature. The competing reactions (model 2) will exhibit stress hardening if aged at lower temperature before stress is increased, whereas the equilibration model (model 1) will not.

The physical model proposed above provides an idea for an alternative approach to plotting data. Suppose that the distribution of thresholds is not uniform, but has a specific value in each device that is independent of stress. Then, although the distribution is not specified by the model, the time transformation between stress levels is specified. In particular, the solution to either of the above differential equations takes the form:

$$A_{2t} = A_{10} \frac{k_1}{k_1+k_2}\left(1-\exp(-(k_1+k_2)t)\right) \qquad (5.10)$$

Assuming device-specific thresholds independent of stress, this means that failure time of device i is the solution in time to

$$\frac{thresh(i)}{A_{10}(i)} = \left(\frac{k_1}{k_1+k_2}\right)\left(1-\exp(-(k_1+k_2)t)\right) \qquad (5.11)$$

For randomly selected devices and large sample sizes, equal values of percent failing correspond very nearly to equal values of $thresh(i)/A_{10}(i)$. Thus, the time transformation plotted by plotting times corresponding to equal percentiles of failure at different stress levels s_1 and s_2 against one another should fit the implicit function:

$$\left(\frac{k_1(s_1)}{k_1(s_1)+k_2(s_1)}\right)\left(1-\exp(-(k_1(s_1)+k_2(s_1))t_1)\right)$$
$$= \left(\frac{k_1(s_2)}{k_1(s_2)+k_2(s_2)}\right)\left(1-\exp(-(k_1(s_2)+k_2(s_2))t_2)\right) \qquad (5.12).$$

For this particular case, this can be expressed explicitly:

$$t_1 = \left(\frac{-1}{(k_1(s_1)+k_2(s_1))}\right) \ln\left(\left(1-\left(\frac{\left(\frac{k_1(s_2)}{k_1(s_2)+k_2(s_2)}\right)}{\left(\frac{k_1(s_1)}{k_1(s_1)+k_2(s_1)}\right)}\right)\times \left(1-\exp(-(k_1(s_2)+k_2(s_2))t_2)\right)\right)\right) \quad (5.13)$$

which is very close to a simple exponential with an asymptote. The functional closeness between Equations 5.13 and 5.14, when

$$\frac{k_1(s_1)}{k_1(s_1)+k_2(s_1)} > \frac{k_1(s_2)}{k_1(s_2)+k_2(s_2)}$$

means that there can be some difficulty in parameter estimation.

$$t_1 = \left(\left(\frac{k_1(s_2)}{k_1(s_1)(k_1(s_2)+k_2(s_2))}\right)\left(1-\exp(-(k_1(s_2)+k_2(s_2))t_2)\right)\right) \quad (5.14)$$

A solution of the time transform (or *acceleration transform*) amounts to determining the parameters k_1 and k_2 as a function of stress, s, which in this case is temperature. Figure 5.4 plots the empirical acceleration transforms from each data set based on times corresponding to equal values of the Kaplan–Meier (Complement 5.7) survival function.

Clearly, the data show some evidence of the kind of curvature one would expect from Equation 5.13 or 5.14. Although it is not definitive, the physical model developed here from first-order rate kinetics principles provides a basis for the UC model developed in Section 5.1.

5.3 Analysis of a Real Data Set

In this section we demonstrate kinetic analysis and experimental design for studying failures in printed wiring boards accelerated by temperature, humidity, and applied voltage. This failure mode has been discussed extensively (Augis et al., 1989; LuValle and Hines, 1992; and Meeker and LuValle, 1995). The data here are a reduced data set, showing the effect of humidity

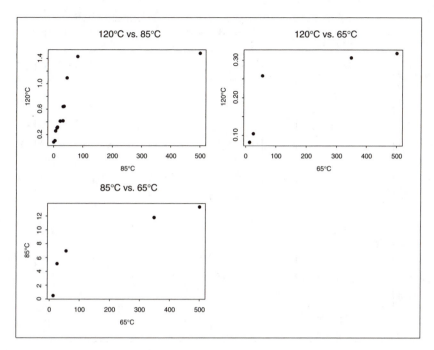

FIGURE 5.4
Acceleration transform plots for each pair of temperature conditions with the first data set.

at fixed high voltage and temperature. The data are interpolated to exact failure times rather than using the original interval censored data. The analysis here is parallel to that shown in Meeker and LuValle (1995). The difference is that here, instead of using a logistic-like transformation, we use a somewhat different approach using acceleration transforms as discussed in Complement 5.5.

The full data set is given in Complement 5.8. It covers experiments run at three levels of humidity while the temperature and applied voltage remain fixed. The first five rows of the data matrix for the highest stress (relative humidity) condition are shown below.

Discounting the row number at the far left, the first column, V1, is the time, the second, V2, is an indicator variable, 1 for failure or 2 if the device was taken off test without a failure occurring. The third column, V3, is temperature (identical for all experiments), and the fourth column, V4, is the function $\left(rh/(1-rh)\right)$, which is the assumed functional form of the dependence of reaction rate on relative humidity (Brauner et al., 1938; Klinger, 1991; Meeker and LuValle, 1995). Relative humidity is measured as a proportion between 0 and 1 throughout.

```
example4a.dat0[[1]]$mat[1:5,]
    V1 V2 V3       V4
1  168  1 85 0.980198
2  431  1 85 0.980198
```

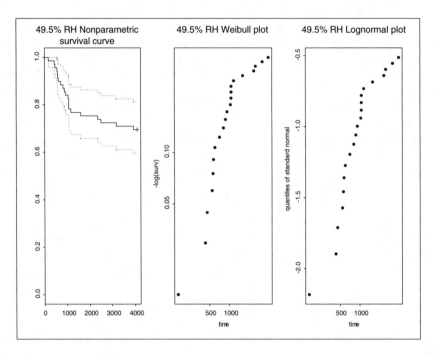

FIGURE 5.5
Nonparametric survival curve, Weibull plot, and lognormal plot for 49.5% relative humidity printed wiring board data.

```
3  459  1  85  0.980198
4  547  1  85  0.980198
5  563  1  85  0.980198
```

The analysis begins with graphical exploration of the failure time distribution.

Tests were run at three levels of relative humidity, 49.5%, 62.8%, and 75.4% consistent with salt solutions available at 85°C. Figure 5.5 contains the data from 49.5% relative humidity, plotted first as a nonparametric survival distribution, $1 - F(t)$, against linear time, then against Weibull and lognormal probability scales, which are transformed to give linear plots with data that fit the respective distributions. The lognormal distribution is one in which the log of the failure time has a normal distribution. It, like the Weibull, is a commonly assumed distribution in reliability analysis.

The first thing we notice from Figure 5.5 is that only about 50% of the devices fail, and neither the Weibull nor the lognormal plot seems to give an especially good straight line. (The linearity of these plots is often used as a goodness-of-fit test for the distribution.) Figure 5.6 shows similar plots for the data from 64.5% relative humidity. Again, only a fraction fail, and neither the lognormal nor the Weibull distribution fits well.

Figure 5.7 shows the survival plots for 75% relative humidity. This time all devices fail, and the plots are considerably more linear. At this point we

Data Analysis for Failure Time Data   151

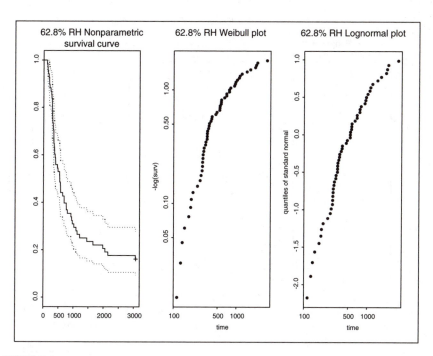

FIGURE 5.6
Nonparametric survival curve, Weibull plot, and lognormal plot for 62.8% relative humidity printed wiring-board data.

need to examine the time acceleration factor between the test conditions of different levels. Recall from Section 5.1 that of the two "standard" models, the one most appropriate for acceleration of a failure mode involving a nonbiological system is a single rate-limiting mechanism is the accelerated failure time model, where distributions at two conditions are linearly related by a time-scaling factor (Equation 5.2). As noted earlier, on Weibull plots this should result in identical curves shifted only on the horizontal axis. The changing curvature between the Weibull plots indicates a nonlinear time transformation.

To better determine if the time transformations between different conditions are linear we plot the acceleration transforms directly based on the survival curves. This results in the plots shown in Figure 5.8.

We see in all three cases that there is a significant departure from linearity. This means the standard model of linear time acceleration does not fit and, therefore, a different one needs to be applied. The curvature in the acceleration transforms is similar to that in our artificial example, so we use the same first-order kinetic models (models 1 and 2 from Section 5.2) although the prime accelerating stress is now relative humidity.

The model we use for relative humidity originated by Brauner et al. (1938) as a model of the adsorption of monolayers at a surface. Klinger (1991) suggested an adaptation of this model to relative humidity acceleration

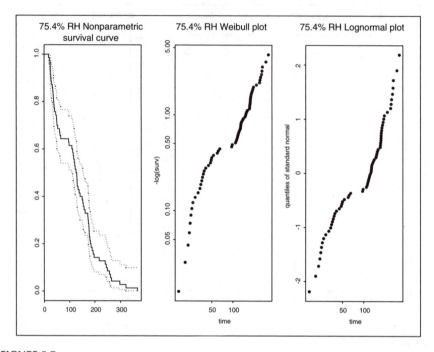

FIGURE 5.7
Nonparametric survival curve, Weibull plot, and lognormal plot for 75.4% relative humidity printed wiring-board data.

effects at interfaces, which is precisely the problem at issue here (LuValle et al., 1986). Step stress experiments (LuValle and Hines, 1992) showed that at the lower relative humidities being studied here the best approximation of the two models was the equilibrium model. Thus, our model of acceleration (ignoring the effects of electric field, as it is constant in these experiments, and assuming the activation energies found in the literature) has the form:

$$A \underset{k_2}{\overset{k_1}{\rightleftarrows}} B$$

$$k_1 = \left(\frac{RH}{1-RH}\right)^{\beta_1} v_1 \exp\left(-\frac{E_{a1}}{kT}\right) \quad (5.15)$$

$$k_2 = \left(\frac{RH}{1-RH}\right)^{\beta_2} v_2 \exp\left(-\frac{E_{a2}}{kT}\right)$$

We assume the activation energy given in the literature of 0.9 eV for the forward direction and 0 for the reverse direction. Otherwise we let the

Data Analysis for Failure Time Data

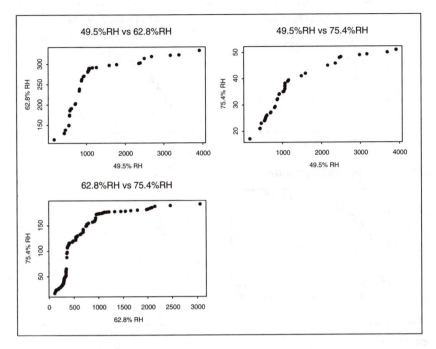

FIGURE 5.8
Acceleration transforms between all pairs of conditions for the printed wiring-board data.

printed wiring-board data determine the coefficients, keeping in mind that bias voltage and temperature are constant while the humidity is set at three different levels. (*Note:* The data for this example are actually a subset from a factorial experiment varying all three stresses; LuValle et al., 1986; LuValle and Mitchell, 1987.)

The analysis presented here was done almost entirely in Splus, using the freeware. Finding good starting values for the acceleration transforms can be a bit tricky. The approach described here is not implemented directly in the freeware and is, in fact, very specific to this particular model.

The approach here is to use each pair of conditions to estimate the approximate transformation given in Equation 5.14 with simplified parameters, i.e.,

$$t_1 = \phi_1(s_1, s_2)\left(1 - \exp\left(-\phi_2(s_2)t_2\right)\right) \tag{5.16}$$

Then we estimate the original rate constants from the set of simplified parameters, using these as starting values in the maximum likelihood estimation. The estimation can be done graphically in either Splus or Excel; the plots below generated in Splus show the results. Figure 5.9 shows the estimated curve with each data set as in Figure 5.8.

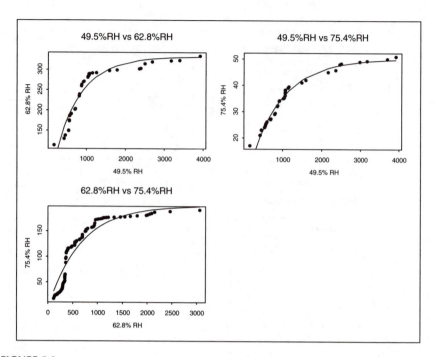

FIGURE 5.9
Least-squares fit of acceleration transforms between all pairs of conditions for the printed wiring board data.

From Equations 5.14 and 5.16, we have that $\phi_2(s_2) = k_1(s_2) + k_2(s_2)$ and $\phi_1(s_1, s_2)\phi_2(s_2) = (k_1(s_2)/k_1(s_1))$. If we postulate that $k_i(s) = \theta_i(T, V)(rh/(1-rh))^{\beta_i}$ then we see that plotting

$$\log\left(\frac{\frac{rh_1}{1-rh_1}}{\frac{rh_2}{1-rh_2}}\right)$$

vs. $\log(\phi_1(s_1, s_2)\phi_2(s_2))$ should give a straight line with intercept 0, and slope β_i.

The values estimated from each plot in Figure 5.9 are

- 49.5% vs. 62.8%, $\phi_1 = 331.98$, $\phi_2 = 0.0015$
- 49.5% vs. 75.4%, $\phi_1 = 50.14$, $\phi_2 = 0.00122$
- 62.8% vs. 75.4%, $\phi_1 = 200.44$, $\phi_2 = 0.00151$

Figure 5.10 shows the plot of the log of the ratio of the relative humidity functions vs. the log of the products of ϕ_1 and ϕ_2 with the least-squares fit

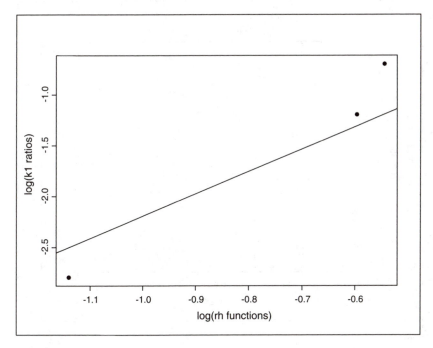

FIGURE 5.10
Plot of $\log\left((rh_1/1-rh_1)/(rh_2/1-rh_2)\right)$ vs. $\log(\phi_1(s_1,s_2)\phi_2(s_2))$ from the two conditions for estimation of the relative humidity power.

forced through the origin added. The slope is approximately 2.2. If we further assume, merely to find a simple starting point for maximum likelihood estimation, that $k_2(s)$ is independent of humidity, then plotting $\phi_2(s_2)$ vs. $\left(rh_2/(1-rh_2)\right)^{2.2}$ should allow us to estimate a line with slope $\theta_1(T,V)$ and intercept $k_2(s_2)$. This plot with a least-squares fit is shown in Figure 5.11. Note that there are two estimates of $\phi_2(s_2)$ at 49.5% relative humidity. The slope is ~6.79e–5, while the intercept is ~1.3e–3.

Direct maximum likelihood analysis using acceleration transforms requires that the distribution that we are converting back to in the base condition not confound the kinetic model. Confounding occurs when the estimation problem of parameters becomes entangled in a way that the individual parameters cannot be uniquely identified. Unfortunately, kinetic models are confounded with any distribution with a scale parameter, and partially confounded with other parameters. To avoid the confounding problem we estimate the distribution using only the data from the base condition (75% relative humidity). Then we estimate the conditional maximum likelihood estimate given that distribution. This turns out to be a particularly simple case of pseudo-maximum likelihood estimation (Gong and Sameniego, 1981). A brief review of the theory is given in Complement 5.6.

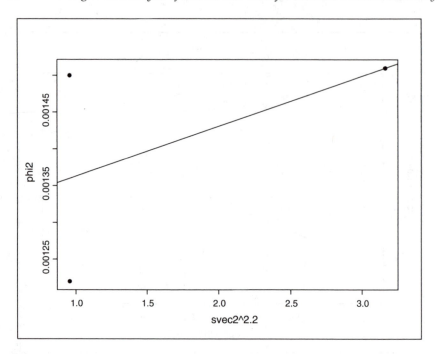

FIGURE 5.11
Plot of ϕ_2 vs. the relative humidity function from the lower stressed condition for the estimation of k_2 and the premultiplier of k_1.

To actually do the pseudo-maximum likelihood estimation using the freeware system, these numbers must be converted from a timescale corresponding to hours to a timescale corresponding to seconds. Thus, we divide by 3600. To be consistent with the literature, we also assume and activation energy for temperature of 0.9 eV. This gives the value of

$$\theta_1(T,V) = \exp(-0.9/kT) * 10^{4.93},$$

so

$$k_1 = \left(rh/(1-rh)\right)^{2.2} \exp(-0.9/kT) * 10^{4.93},$$

and $k_2 = 3.6e - 7$.

Our initial estimates of the Weibull parameters at the 75.4% relative humidity condition are 140.4235 for the scale, and 1.54 for the shape (the likelihood was −400.362).

The fit to the raw data for these values of the parameters and the maximum likelihood estimate for a Weibull distribution at 75.4% relative humidity is shown as the plot on the left in Figure 5.12. Pseudo- maximum likelihood estimates result in the plot on the right of Figure 5.12.

The estimated parameter values are

Data Analysis for Failure Time Data

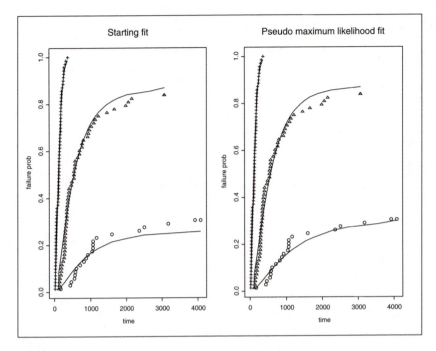

FIGURE 5.12
Plot of fit of initial parameters and pseudo-maximum likelihood estimates to the raw failure time distributions.

$$k_1 = 10^{5.46} \exp\left(-\frac{0.9}{kT}\right)\left(\frac{rh}{1-rh}\right)^{2.38} \quad (5.17)$$

$$k_2 = 10^{-6.62}$$

To predict the probability of failure at any given condition, we just calculate the equivalent time under our model for the base condition, 75.4% relative humidity, and find the Weibull probability under the estimated failure time distribution at 85°C, 75.4% relative humidity. For simplicity we assume the same high voltage as was used in the experiment. Thus, at 40°C 50% relative humidity, the acceleration transform is calculated by substituting the values of Equation 5.17 into Equation 5.13, with s_1 = 85°C, 75.4% relative humidity, and s_2 = 40°C 50% relative humidity. Substituting these values, for 20 years = 630,720,000 s, we find that the equivalent time at 85°C, 75.4% relative humidity is

$k_1(40C, 50\%RH) \approx 9.62e-10$

$k_1(85C, 75\%RH) \approx 9.12e-7$

$k_2 \approx 2.39e-7$

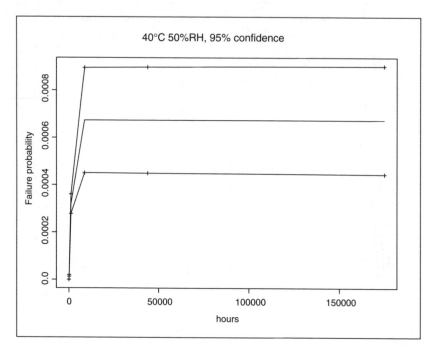

FIGURE 5.13
Extrapolation of failure probability to 20 years at 40°C, 50% relative humidity with 95% confidence bounds.

$$t_{85C,75\%RH} = \left(\frac{-1}{9.12e-7+2.39e-7}\right) \ln\left(1-\left(\frac{(9.62e-10)(9.12e-7+2.39e-7)}{(9.12e-7)(9.62e-10+2.39e-7)}\right)\left(1-\exp\left(\frac{-(9.62e-10+2.39e-7)}{\times 20\text{yrs}}\right)\right)\right)$$

By substituting the number of seconds, this is approximately 1.22 h. Combining this with the estimated Weibull distribution (scale = 140.42, shape = 1.54) this gives approximately 6.7e–4 as the probability of failure during life.

Confidence bounds can now be calculated based on the pseudo-maximum likelihood theory given in Complement 5.6. The 20-year prediction for cumulative failure probability is given in Figure 5.13, with 95% confidence bounds.

5.3.1 Summary

In this chapter approaches to integrating physical reasoning into the analysis of accelerated tests have been studied using two examples. Both approaches

Data Analysis for Failure Time Data

involve postulating unobserved processes with physical explanations that give rise to data consistent with the observed data. The first is based on assuming stochastic processes (a failure process X(t) and a UC process Y(t)), whereas the second postulates chemical kinetics as the primary driver changing the way aging occurs at different conditions. The latter is preferable because it provides more power to generate falsifiable experiments (e.g., step stress experiments, identification of species involved), but is often more difficult to develop analytically. There is a theoretical framework, statistical kinetics (LuValle et al., 1988), which unifies the two approaches through theory related to the so-called master equation (Oppenheim et al., 1977).

Although the statistical kinetic theory allows direct maximum likelihood, it corresponds to a rather restrictive assumption of a uniform distribution of thresholds. Thus, the acceleration transform methodology, while requiring pseudo-maximum likelihood estimation, should be more general.

One issue we have not covered in this chapter is the issue of designing experiments to improve the estimates, or to test for differences between statistical models. Complement 5.10 suggests an exercise for readers that allows them to use the software plus the material from earlier chapters to look at some of these more classical questions with designs more robust to model assumptions.

5.4 Complement: Maximum Likelihood Analysis

Maximum likelihood analysis can be used to choose the most reasonable model, or the most plausible parameter values for a given model, to fit a data set. It is based on finding parameter values that maximize the *likelihood* of the observed sample outcome. The likelihood of a parameter value, given a model and data set, is (almost) proportional to the probability of the data given the model. With continuous distributions like the Weibull and the exponential, the probability of failure at any single point in time is 0, so we take the likelihood as the value of the probability density at the observed data point. Assuming each data point is independent, the likelihood of a parameter value, θ, given a data set with observed failures (in set D) and censored data (in set C) would take the form:

$$l(\bar{x},\theta) = \prod_{i \in D} f(x_i,\theta) \prod_{j \in C} (1 - F(x_j,\theta)) \qquad (5.18)$$

For convenience, both numerical and mathematical, it is often easier to deal with the natural log of this term:

$$L(\bar{x},\theta) = \sum_{i \in D} \ln(f(x_i,\theta)) + \sum_{j \in C} \ln((1 - F(x_j,\theta))) \qquad (5.19)$$

Assume for simplicity that the experiment is run simultaneously on all devices and ends at a point in time t_c. Then the log likelihood can be written:

$$\sum_{i=1}^{n} \ln(g(x_i,\theta)) \qquad (5.20)$$

where

$$g(x_i,\theta) = \begin{cases} f(x_i,\theta) & \text{if } x_i < t_c \\ 1-F(t_c,\theta) & \text{otherwise} \end{cases} \qquad (5.21)$$

For the sake of completeness, some results that can be found in a number of elementary texts in mathematical statistics are provided here. As a starting point we assume the law of large numbers and the central limit theorem (e.g., Serfling, 1980).

5.4.1   Law of Large Numbers

Let $X_i, i = 1,...,n$ be a set of independent and identically distributed random variables with distribution F. Let $\int X dF = E(X) = \mu$ exist. Then the term

$$\overline{X}_n = \frac{\sum_{i=1}^{n} X_i}{n}$$

converges to μ in probability as $n \to \infty$, where convergence in probability of a sequence of random variables Y_n to a random variable Y is defined by: $Y_n \xrightarrow{P} Y$ if and only if, for all $\varepsilon > 0, \delta > 0$ there exists N such that for all $n > N$, $P(|Y_n - Y| > \varepsilon) < \delta$. In fact, the law of large numbers can be stated in a much stronger fashion, but in this book, to avoid making the text too long, the mathematical theory is presented more at the level of rigor that engineers are accustomed to. Those desiring more rigorous treatment are welcome to review the references.

5.4.2   Central Limit Theorem

Let $x_i, i = 1,...,n$ be a realization of a set of independent and identically distributed random variables with distribution F. Let

$$\int X dF = E(X) = \mu \text{ exist,}$$

and suppose also that

$$\int (X - \mu)^2 dF = \sigma^2 < \infty.$$

Then the term

$$n^{1/2}(\bar{x}_n - \mu)$$

converges in distribution to the law of a normal random variable with mean 0 and variance σ^2. This means that no matter the form of F, as long as it obeys the above assumptions, the distribution of $n^{1/2}(\bar{x}_n - \mu)$ converges to the distribution with density:

$$f(x) = \frac{1}{\sqrt{2\pi\sigma^2}} \exp\left(-\left(\frac{x^2}{2\sigma^2}\right)\right)$$

5.4.3 Proof of Consistency of Maximum Likelihood

OBSERVATION
Define $k(z) = \ln(z) - z$. The derivative of k is $(1/z) - 1$ and is 0 only at $z = 1$. It can be seen by inspection that this corresponds to the maximum value $k(z)$, which is -1. From this we develop the following results. We assume that both $f(x)$ and $g(x)$ are generalized density functions and hence integrate to 1.

We note that the first two lines below hold for each x; hence the last line holds, with equality only when $f(x) \equiv g(x)$.

$$\ln\left(\frac{f(x)}{g(x)}\right) \leq \left(\frac{f(x)}{g(x)}\right) - 1 \Rightarrow$$

$$\ln\left(\frac{f(x)}{g(x)}\right) g(x) \leq (f(x) - g(x)) \Rightarrow \quad (5.22)$$

$$\int \ln\left(\frac{f(x)}{g(x)}\right) g(x) dx \leq \int (f(x) - g(x)) dx \equiv 0$$

The last line above is equivalent to

$$\int \ln(f(x)) g(x) dx \leq \int \ln(g(x)) g(x) dx \quad (5.23)$$

For the case of time-censored data, a similar result can be derived by replacing the integral sign with the sum of an integral and a discrete term corresponding to the censoring time.

We now use Equation 5.23 to show an important property of the log likelihood from Expression 5.20. Since the x_i are realizations of independent and identically distributed random variables, each term $g(x_i, \theta)$ is as well. Thus, by the law of large numbers, for each $\tilde{\theta}$,

$$\frac{\sum_{i=1}^{n} \ln(g(x_i, \tilde{\theta}))}{n} \xrightarrow{n \to \infty} \int \ln(g(x, \tilde{\theta})) g(x, \theta) dx \qquad (5.24)$$

where we assume that $g(x,\theta)$ is the true model. Because the left side of Equation 5.24 is maximized by $\tilde{\theta} = \theta$ (as shown by Equation 5.23), it follows that the maximum value of the log likelihood estimator is consistent when the class of models being investigated includes the true model. Rigorous proof requires a proof of uniform convergence of Equation 5.24, which follows from sufficient smoothness conditions on $g(x,\theta)$. Alternative proofs of both consistency and efficiency can be found in, for example, Serfling (1980) or Cramer (1946).

5.4.4 Derivation of the Distribution of the Maximum Likelihood Estimator

The distribution can be derived through the use of Taylor expansions. In particular, if θ_0 is the true value and $\hat{\theta}$ the maximum likelihood estimator, then

$$\sum_{i=1}^{n} \frac{\partial \ln(g(x_i, \theta_0))}{\partial \theta} = \sum_{i=1}^{n} \frac{\partial \ln(g(x_i, \theta_0))}{\partial \theta} - \sum_{i=1}^{n} \frac{\partial \ln(g(x_i, \hat{\theta}))}{\partial \theta}$$

$$\approx (\theta_0 - \hat{\theta}) \left[\frac{\partial^2}{\partial \theta_j \partial \theta_k} \sum_{i=1}^{n} \ln(g(x_i, \hat{\theta})) \right] \qquad (5.25)$$

Thus,

$$(\hat{\theta} - \theta_0) = -\left[\frac{\partial^2}{\partial \theta_j \partial \theta_k} \sum_{i=1}^{n} \ln(g(x_i, \theta)) \right]^{-1} \left(\sum_{i=1}^{n} \frac{\partial}{\partial \theta} \ln(g(x_i, \theta_0)) \right) \qquad (5.26)$$

To proceed from here, multiply both sides of Equation 5.26 by $\sqrt{n}$, and rewrite it as

$$\sqrt{n}(\hat{\theta} - \theta_0) = \sqrt{n} \left[-\frac{\sum_{i=1}^{n} \frac{\partial^2}{\partial \theta_j \partial \theta_k} \ln(g(x_i, \theta_0))}{n} \right]^{-1} \left(\frac{\sum_{i=1}^{n} \frac{\partial}{\partial \theta} \ln(g(x_i, \theta))}{n} \right) \qquad (5.27)$$

The first term on the left-hand side (within the square brackets) converges by the law of large numbers and continuity of the matrix inverse to

$$-E\left(\frac{\partial^2}{\partial \theta_j \partial \theta_k} \ln(g(X,\theta))\right)^{-1} \quad (5.28)$$

The second term is a sum of independent and identically distributed random variables, which if it has finite variance (actually covariance matrix, as the derivative gives a vector) and 0 mean, will result in a Gaussian distribution. To finish our indication of the maximum likelihood theory, we need two identities. To obtain these identities we will be a bit more careful, because of the general way we are defining distributions conditional on our experiment. We follow Serfling here.

$$\int \frac{\partial g(x,\theta)}{\partial \theta} = \int_0^c \frac{\partial f(x,\theta)}{\partial \theta} dx + \frac{\partial(1 - F(C,\theta))}{\partial \theta}$$

$$= \frac{\partial}{\partial \theta}\left(\int_0^C f(x,\theta)dx + (1 - F(C,\theta))\right) = \frac{\partial}{\partial \theta}(1) = 0$$

where the second equality follows from the fact that C is independent of θ and f is equicontinuous. Thus,

$$\int \frac{\partial^2}{\partial \theta_j \partial \theta_k} g(x,\theta) = 0$$

as well. For simplicity in notation, we assume a scalar θ below.

$$E\left(\frac{\partial \ln(g(x,\theta))}{\partial \theta}\right) = \int \frac{1}{g(x,\theta)} \frac{\partial g(x,\theta)}{\partial \theta} g(x,\theta) dx = 0 \quad (5.29)$$

and

$$E\left(\frac{\partial^2 \ln(g(x,\theta))}{\partial \theta^2}\right) = \int \left(\frac{1}{g(x,\theta)} \frac{\partial^2}{\partial \theta^2} g(x,\theta) - \frac{\left(\left(\partial/\partial\theta\right)g(x,\theta)\right)^2}{g(x,\theta)}\right) dx$$

$$= -E\left(\left(\frac{\partial \ln(g(x,\theta))}{\partial \theta}\right)^2\right)$$

$$(5.30)$$

Thus, the last term on the left side of Equation 5.27 has mean 0 and covariance matrix Σ_I, the inverse of Equation 5.28. So we find that the right side of Equation 5.27 converges to a Gaussian random vector with mean 0 and covariance $\Sigma_I^{-1}\Sigma_I\Sigma_I^{-1} = \Sigma_I^{-1}$, where Σ_I is known in statistics as the information matrix.

We note that the maximum likelihood estimator can be found in several ways. In the text, we either solve the derivative of the log likelihood directly when possible or use numerical optimization of the log likelihood.

Suppose that we have one model (with parameter vector θ) that is a special case of another model (with parameter vector ϕ), for example, the exponential model is a special case of the Weibull model, with $\beta = 1$; then $-2(L(\hat{\theta}) - L(\hat{\phi}))$ has a chi square distribution with the number of degrees of freedom equal to the difference in the number of estimated parameters. There are some difficulties in using this formulation for comparing the exponential model to the UC model, because although the exponential model is a submodel, it is one that lies on an edge of the parameter space. However, we will assume that we conservatively estimate the sampling distribution of $-2(L(\hat{\theta}) - L(\hat{\phi}))$ by the sampling distribution of $-2(L(\theta_0) - L(\hat{\phi}))$, so the degrees of freedom will be the number of parameters in ϕ.

Maximum likelihood parameter estimates have minimum variance among other possible estimators for large samples. They also are invariant under transformation, so the maximum likelihood estimate of a function $g(\theta)$ where θ is the parameter being estimated is just that function of the maximum likelihood estimator. Predictions based on maximum likelihood estimates are functions of the estimator, so the asymptotic covariance of a vector prediction $\vec{g}(\hat{\theta}) = (g_1(\hat{\theta}), \ldots, g_k(\hat{\theta}))$ is just

$$\left(\frac{\partial}{\partial \theta}\vec{g}(\theta)\right)' \Sigma_I^{-1} \left(\frac{\partial}{\partial \theta}\vec{g}(\theta)\right) \tag{5.31}$$

5.4.5 Splus Source Code

Below is the source code for Splus for the analysis of the first (artificial) data set in Chapter 5

```
THIS PROGRAM CALCULATES THE LOG LIKELIHOOD FOR A SINGLE DATA SET UNDER
THE UNOBSERVABLE CENSORING MODEL GIVEN k1(= λ₁) and k2(= λ₂)
"unobs.cens.lik"<-
function(mat, k1, k2)
{
        cens <- mat[, 2]
        I1 <- as.logical(cens)
        I2 <- as.logical(1 - cens)
        I3 <- 1 - cens
        fail <- mat[I1, 1]
        nof <- mat[I2, 1]
        lik.fail <- length(fail) * log(k1) - (k1 + k2) * sum(fail)
        lik.cens <- 0
        if(sum(I3) > 0.5) {
```

```
            P1 <- k1/(k1 + k2)
            dum <- log(P1 * exp(- (k1 + k2) * nof) + (1 - P1))
            lik.cens <- sum(dum)
    }
    lik.fail + lik.cens
}
```

THIS PROGRAM CALCULATES THE LOG LIKELIHOOD FOR A SINGLE DATA SET UNDER THE EXPONENTIAL MODEL GIVEN k1(= λ_1) and k2(= λ_2)

```
"exp.lik"<-
function(mat)
{
    cens <- mat[, 2]
    fail <- mat[, 1]
    lambda <- sum(cens)/sum(fail)
    lik <- sum(cens) * log(lambda) - lambda * sum(fail)
    list(lik = lik, lambda = lambda)
}
```

THIS PROGRAM CALCULATES THE negative JOINT LOG LIKELIHOOD FOR ALL THE DATA SETS UNDER THE UNOBSERVABLE CENSORING MODEL GIVEN A PARAMETER VECTOR $\left[v_1, E_{a1}, v_2, E_{a2}\right]$

```
"unobs.cens.lik.regss"<-
function(parvec, dat = unobs.cens.data1)
{
    n1 <- length(dat)
    llik <- 0
    for(i in 1:n1) {
        m1 <- dat[[i]]
        temp <- mean(m1[, 3])
        kT <- 1/(boltzmann * (temp + 273))
        k1 <- 10^(parvec[1]) * exp(- parvec[2] * kT)
        k2 <- 10^(parvec[3]) * exp(- parvec[4] * kT)
        llik <- llik + unobs.cens.lik(m1, k1, k2)
    }
    list(ss0 = -llik)
}
```

THIS PROGRAM CALCULATES THE negative JOINT LOG LIKELIHOOD FOR ALL THE DATA SETS UNDER THE EXPONENTIAL MODEL GIVEN A PARAMETER VECTOR $\left[v_0, E_{a0}\right]$

```
"exp.lik.regss"<-
function(parvec, dat = unobs.cens.data1)
{
    n1 <- length(dat)
    llik <- 0
    for(i in 1:n1) {
        m1 <- dat[[i]]
        temp <- mean(m1[, 3])
        kT <- 1/(boltzmann * (temp + 273))
        k1 <- 10^(parvec[1]) * exp(- parvec[2] * kT)
        llik <- llik + exp.lik0(m1, k1)
    }
```

```
        llik
        list(ss0 = -llik)
}
```

THIS IS THE VECTOR OF OPTIMIZED VALUES FOR THE UNOBSERVABLE CENSORING MODEL

```
"uc.parvec0"<-
c(16.57518784492766, 1.29, 1.491361693834273, 0.25)
```

THIS IS THE VECTOR OF OPTIMIZED VALUES FOR THE EXPONENTIAL MODEL

```
"exp.parvec0"<-
c(20.11394335230684, 1.57)
```

THIS PROGRAM CALCULATES THE LOG LIKELIHOOD FOR A SINGLE DATA SET UNDER THE EXPONENTIAL MODEL GIVEN k1 (= λ_1) and k2 (= λ_2)

```
"exp.lik0"<-
function(mat, k1)
{
        cens <- mat[, 2]
        fail <- mat[, 1]
        lambda <- k1
        lik <- sum(cens) * log(lambda) - lambda * sum(fail)
        lik
}
```

5.5 Complement: Statistical Estimation of Kinetics from Failure Time Data

In this section we develop the statistical theory that allows estimation of kinetic models from failure time data.

The physical model we use starts off just as the kinetic models in Chapter 1:

$$\frac{dA_t}{dt} = K(T)A_t \tag{5.32}$$

with an internal degradation of the form:

$$Y_{it} = c_\lambda A_{it} \tag{5.33}$$

where c_λ is a fixed vector resulting in a linear combination of the states involved in the degradation process that ultimately leads to failure if the exposure to stress continues long enough. The translation from the model to actual data comes by assuming that the observable (failure time) is the

Data Analysis for Failure Time Data

time when the value of Y_{it} for device i exceeds a threshold C_i. The essential feature of a statistical model is that this threshold value is not the same for all devices; i.e., it has a distribution. It may or may not correspond to actual failure, depending on the engineering details, but it produces the same observable response regardless of the value of C_i. We assume that the threshold is fixed for each device, and *does not vary with stress*. This latter assumption has to be thought about carefully. In some cases it may simply be wrong in which case a more complex measurement scheme may be necessary.

A subset of kinetic models results in a monotonic increase of Y_{it} over time, for any stress pattern. For these models, an acceleration transform (LuValle et al., 1986, 1988) can be uniquely defined between two stress conditions. For example, for the simple case of temperature stress, with T_1 and T_2 as the two temperature conditions, then the acceleration transform τ:

$$t_1 = \tau(t_2, T_2, T_1, \theta) \tag{5.34}$$

where time t_1 under temperature 1 is related to time t_2 under temperature 2 by the implicit equation:

$$Y_{iT_1 t_1} = Y_{iT_2 t_2} \tag{5.35}$$

and θ is the vector of parameters specific to the kinetic equation representing the degradation reaction.

Assuming that C_i corresponds to failure and the devices are randomized across conditions so the distributions at each condition are close to one another we can estimate τ by estimating the time transformation that satisfies the relation:

$$G(T_1, t_1) = G(T_2, \tau(t_2))$$

where $G(T, t)$ is the cumulative failure distribution at temperature T at time t (LuValle, 1999). This is equivalent to saying that the distribution of *times* to reach threshold values shifts and distorts with changing stress, but each device maintains its position (percentile) in the time order of failure.

There are several ways to formulate the statistical problem of estimating an acceleration transform (e.g., Meeker and LuValle, 1995). Here we take a particularly simple approach. We assume that at one stress condition the observed failure distribution can be well approximated with particular failure distribution. For discussion we use the Weibull simply because it has proved useful, but it could be any one of a number of other distributions. Explicitly, the Weibull distribution has the cumulative distribution function:

$$G(T, t) = W(t; \alpha, \beta) = P(t < t_0) = 1 - \exp\left(-\left(\frac{t_0}{\alpha}\right)^\beta\right) \tag{5.36}$$

If the temperature for which the Weibull holds is denoted T_1, the distribution at any other temperature T_2 has the form:

$$G(T_2, t_2; \alpha, \beta, \theta) = W(\tau(t_2, T_2, T_1, \theta), \alpha, \beta) \tag{5.37}$$

and the probability density for absolutely continuous distributions (such as the Weibull) has the form:

$$g(T_2, t_2; \alpha, \beta, \theta) = \frac{d}{dt_2}(\tau(t_2, T_2, T_1, \theta)) w(\tau(t_2, T_2, T_1, \theta); \alpha, \beta) \tag{5.38}$$

We use maximum likelihood to estimate the parameters of the Weibull distribution as described in Complement 5.4 from the high-stress condition, and pseudo-maximum likelihood to estimate the kinetic time transformation, assuming the Weibull parameters. Reasons for maximum likelihood can be found in a number of references (Serfling, 1980; Nelson, 1982; Lawless, 1982; Meeker and Escobar, 1998). If we denote the parameters for the Weibull distribution as α and β. The likelihood for k sets of data across several conditions is evaluated as

$$L = \prod_{j=1}^{k} \left(\left(\prod_{i \in M_j} g(T_j, t_{ji}; \alpha, \beta, \theta) \right) \prod_{i \in C_j} 1 - G(T_j, t_{ji}; \alpha, \beta, \theta) \right) \tag{5.39}$$

where M_j indicates the set of objects that fails during the experiment, and C_j indicate the objects that do not fail before the experiment is terminated (the times then represent time under test, and are called censoring times). For numerical and statistical reasons (e.g., Serfling, ibid) it is easier to work with the log of Expression 5.39. In the freeware we minimize the term $-\log(L)$, which is equivalent to maximizing the pseudo likelihood. To allow for the uncertainty in starting times the pseudo likelihood may have the form:

$$L = \prod_{j=1}^{k} \left(\left(\prod_{i \in F_j} g(T_j, t_{ji} + \Delta_j; \alpha, \beta, \theta) \right) \prod_{i \in M_j} 1 - G(T_j, t_{ji} + \Delta_j; \alpha, \beta, \theta) \right) \tag{5.40}$$

We assume here that the time uncertainty can be associated with each experimental cell separately. If we define $\phi = \{\Delta_j, j = 1, ..., k, \theta\}$ and if we define $\vec{t}$ as the vector of all failure and censoring times from the experiments, then the statistical problem is to determine the behavior of the estimate $\hat{\phi}$ that maximizes:

$$l(\hat{\phi}, \vec{t}) = \log(L(\hat{\phi}, \vec{t})) \tag{5.41}$$

Data Analysis for Failure Time Data

The form of Equation 5.41 is that of a sum of independent, bounded, random variables. Thus, for any given value of ϕ, the central limit theorem (Serfling, 1980) implies that appropriately normalized $l(\hat{\phi}, \vec{t})$ will converge to a Gaussian distribution. With a certain amount of mathematics, it turns out that $(\hat{\phi} - \phi)$, if α and β are known (where $\hat{\phi}$ is the maximum likelihood estimate and ϕ is the true value) and appropriately normalized, will also converge to a Gaussian distribution, with mean 0 and covariance matrix:

$$\Sigma_{\hat{\phi}} = E\left[\left(\frac{dl(\phi, \vec{t})}{d\phi}\right)\left(\frac{dl(\phi, \vec{t})}{d\phi}\right)'\right]^{-1} = E\left[\left(\frac{d^2l(\phi, \vec{t})}{d\phi^2}\right)\right]^{-1} \quad (5.42)$$

as shown in Complement 5.4. In the freeware program, used for calculations in much of this book, the covariance estimate of $(\hat{\phi} - \phi)$ is based on using the empirical estimate of the left-hand side of Equation 5.42.

5.6 Complement: Pseudo-Maximum Likelihood Estimation

Unfortunately, the maximum likelihood theory assumes that the parameters can all be estimated from the data in a consistent way. The Weibull distribution and many kinetic models are at least partially confounded. Thus, what we do is condition on the Weibull parameters estimated from the base condition for the acceleration transform, and then estimate the parameters of the kinetics, assuming those are fixed. This is called pseudo-maximum likelihood estimation (studied in more generality by Gong and Sameniego, 1981) and takes a particularly simple form here. The base condition only enters into the estimates in that it sets the Weibull distribution. Then the other conditions apply to the estimation of the kinetic parameters.

To denote the result, divide ϕ into $(\phi_1, \phi_2) = (\phi_1, \alpha, \beta)$; then define

$$I = E\left[\left(\frac{dl(\phi, \vec{t})}{d\phi}\right)\left(\frac{dl(\phi, \vec{t})}{d\phi}\right)'\right]$$

$$= E\left[\begin{pmatrix} \frac{dl(\phi, \vec{t})}{d\phi_1} \\ \frac{dl(\phi, \vec{t})}{d\phi_2} \end{pmatrix} \left(\frac{dl(\phi, \vec{t})}{d\phi_1}, \frac{dl(\phi, \vec{t})}{d\phi_2}\right)\right] = \begin{pmatrix} I_{11} & I_{12} \\ I_{21} & I_{22} \end{pmatrix} \quad (5.43)$$

Define $\tilde{\phi}_2 = (\tilde{\alpha}, \tilde{\beta})'$ as the maximum likelihood estimates based only on the base condition and $\hat{\phi}_1$ as the maximum likelihood estimate of ϕ_1 conditioned on $\tilde{\phi}_2$. Following Gong and Sameniego, $\sqrt{(n)}(\hat{\phi}_1 - \phi_1)$ is asymptotically equivalent to

$$\left(\sqrt{(n)} \left(\frac{d}{d\phi_1} l(\phi, t) \right) - \sqrt{(n)}(\tilde{\phi}_2 - \phi_2) I_{12} \right) I_{11}^{-1}$$

Here we use $\sqrt{(n)}$ as a notation to imply we are doing the correct normalization so the variable converges in law to a nontrivial Gaussian distribution. Thus, the asymptotic (large sample) covariance of

$$\sqrt{(n)}(\hat{\phi}_1 - \phi_1)$$

has the form shown in Equation 5.44 below. The two expectations in the middle vanish because the Weibull parameters are only estimated at the base condition, but the kinetic transformation at the base condition is always the identify function; thus, the two terms in the asymptotic representation of the pseudo likelihood are statistically independent.

$$E \begin{bmatrix} I_{11}^{-1} \left(\left(\sqrt{(n)} \frac{d}{d\phi_1} l(\phi,t) \right)' - \left(\sqrt{(n)} I_{12} '(\tilde{\phi}_2 - \phi_2) \right)' \right) \\ \left[\left(\sqrt{(n)} \frac{d}{d\phi_1} l(\phi,t) \right) - \left(\sqrt{(n)} I_{12}(\tilde{\phi}_2 - \phi_2) \right) \right] I_{11}^{-1} \end{bmatrix}$$

$$= I_{11}^{-1} I_{11} I_{11}^{-1} + I_{11}^{-1} I_{12}' \Sigma(\tilde{\phi}_2) I_{12} I_{11}^{-1}$$

$$- E \left[\left(\sqrt{(n)} \frac{d}{d\phi_1} l(\phi,t) \right) \left(\sqrt{(n)} I_{12}'(\tilde{\phi}_2 - \phi_2) \right) \right] \qquad (5.44)$$

$$- E \left[\left(\sqrt{(n)} I_{12}'(\tilde{\phi}_2 - \phi_2) \right) \left(\sqrt{(n)} \frac{d}{d\phi_1} l(\phi,t) \right) \right]$$

$$= I_{11}^{-1} + I_{11}^{-1} I_{12}' \Sigma(\tilde{\phi}_2) I_{12} I_{11}^{-1}$$

5.7 Complement: The Kaplan–Meier Estimate

The Kaplan–Meier estimate of the survival function using censored data is an estimate of the failure distribution that is not dependent on parametric statistical assumptions (e.g., Kabfliesch and Prentice, 1980). Thus, it is good for calculating equivalent quantiles when constructing the acceleration transform graphically. Suppose n devices are put on test at time 0, and observed continuously. Suppose that they are observed continuously, but periodically, some devices are needed for some other purpose, and survivors are pulled off of test. The probability of failure must be estimated conditionally based on the number of devices on test when failure is observed. The Kaplan–Meier estimate is structured to accomplish this as follows.

Let $j = 1, \ldots J$ index the times where devices either fail or are pulled off test without failing (censored). Then the probability of failure by time t is estimated as

$$1 - \hat{S}(t); \quad \hat{S}(t) = \prod_{\{j: t_j < t\}} \left(\frac{n_j - m_j}{n_j} \right) \qquad (5.45)$$

where n_j is the number on test at time t_j, and m_j is the number failing at time t_j.

5.8 Complement: Printed Wiring-Board Data

In this section the printed wiring-board failure data, which are used in the second example, are provided. Explanation of the data columns is given in Section 5.3.

49.5% Relative Humidity

```
1   168    1 85 0.980198
2   431    1 85 0.980198
3   459    1 85 0.980198
4   547    1 85 0.980198
5   563    1 85 0.980198
6   575    1 85 0.980198
7   603    1 85 0.980198
8   706    1 85 0.980198
9   814    1 85 0.980198
10  874    1 85 0.980198
11  922    1 85 0.980198
12 1040    1 85 0.980198
13 1064    1 85 0.980198
14 1065    1 85 0.980198
```

```
15  1066  1  85  0.980198
16  1160  1  85  0.980198
17  1592  1  85  0.980198
18  2359  1  85  0.980198
19  2504  1  85  0.980198
20  3176  1  85  0.980198
21  3920  1  85  0.980198
22  4078  2  85  0.980198
23  4078  2  85  0.980198
24  4078  2  85  0.980198
25  4078  2  85  0.980198
26  4078  2  85  0.980198
27  4078  2  85  0.980198
28  4078  2  85  0.980198
29  4078  2  85  0.980198
30  4078  2  85  0.980198
31  4078  2  85  0.980198
32  4078  2  85  0.980198
33  4078  2  85  0.980198
34  4078  2  85  0.980198
35  4078  2  85  0.980198
36  4078  2  85  0.980198
37  4078  2  85  0.980198
38  4078  2  85  0.980198
39  4078  2  85  0.980198
40  4078  2  85  0.980198
41  4078  2  85  0.980198
42  4078  2  85  0.980198
43  4078  2  85  0.980198
44  4078  2  85  0.980198
45  4078  2  85  0.980198
46  4078  2  85  0.980198
47  4078  2  85  0.980198
48  4078  2  85  0.980198
49  4078  2  85  0.980198
50  4078  2  85  0.980198
51  4078  2  85  0.980198
52  4078  2  85  0.980198
53  4078  2  85  0.980198
54  4078  2  85  0.980198
55  4078  2  85  0.980198
56  4078  2  85  0.980198
57  4078  2  85  0.980198
58  4078  2  85  0.980198
59  4078  2  85  0.980198
60  4078  2  85  0.980198
61  4078  2  85  0.980198
62  4078  2  85  0.980198
63  4078  2  85  0.980198
64  4078  2  85  0.980198
65  4078  2  85  0.980198
66  4078  2  85  0.980198
67  4078  2  85  0.980198
68  4078  2  85  0.980198
69  4078  2  85  0.98019
```

64.5% Relative Humidity

```
 1   114  1  85  1.688172
 2   130  1  85  1.688172
 3   138  1  85  1.688172
 4   150  1  85  1.688172
 5   175  1  85  1.688172
 6   187  1  85  1.688172
 7   191  1  85  1.688172
 8   203  1  85  1.688172
 9   239  1  85  1.688172
10   263  1  85  1.688172
11   271  1  85  1.688172
12   283  1  85  1.688172
13   286  1  85  1.688172
14   287  1  85  1.688172
15   290  1  85  1.688172
16   292  1  85  1.688172
17   299  1  85  1.688172
18   303  1  85  1.688172
19   319  1  85  1.688172
20   323  1  85  1.688172
21   338  1  85  1.688172
22   339  1  85  1.688172
23   340  1  85  1.688172
24   343  1  85  1.688172
25   355  1  85  1.688172
26   363  1  85  1.688172
27   367  1  85  1.688172
28   387  1  85  1.688172
29   399  1  85  1.688172
30   403  1  85  1.688172
31   463  1  85  1.688172
32   483  1  85  1.688172
33   535  1  85  1.688172
34   538  1  85  1.688172
35   540  1  85  1.688172
36   563  1  85  1.688172
37   567  1  85  1.688172
38   571  1  85  1.688172
39   639  1  85  1.688172
40   689  1  85  1.688172
41   691  1  85  1.688172
42   761  1  85  1.688172
43   763  1  85  1.688172
44   798  1  85  1.688172
45   918  1  85  1.688172
46   942  1  85  1.688172
47   966  1  85  1.688172
48  1020  1  85  1.688172
49  1092  1  85  1.688172
50  1128  1  85  1.688172
51  1212  1  85  1.688172
52  1452  1  85  1.688172
53  1656  1  85  1.688172
```

```
54  1980     1  85  1.688172
55  2052     1  85  1.688172
56  2148     1  85  1.688172
57  3060     1  85  1.688172
58  3067     2  85  1.688172
59  3067     2  85  1.688172
60  3067     2  85  1.688172
61  3067     2  85  1.688172
62  3067     2  85  1.688172
63  3067     2  85  1.688172
64  3067     2  85  1.688172
65  3067     2  85  1.688172
66  3067     2  85  1.688172
67  3067     2  85  1.688172
68  3067     2  85  1.688172
```

85% Relative Humidity

```
 1   17.0    1  85  3.065041
 2   21.0    1  85  3.065041
 3   23.0    1  85  3.065041
 4   24.0    1  85  3.065041
 5   24.7    1  85  3.065041
 6   25.3    1  85  3.065041
 7   26.0    1  85  3.065041
 8   27.0    1  85  3.065041
 9   29.0    1  85  3.065041
10   32.0    1  85  3.065041
11   34.0    1  85  3.065041
12   35.0    1  85  3.065041
13   36.0    1  85  3.065041
14   37.0    1  85  3.065041
15   38.0    1  85  3.065041
16   39.0    1  85  3.065041
17   41.0    1  85  3.065041
18   45.0    1  85  3.065041
19   48.0    1  85  3.065041
20   49.0    1  85  3.065041
21   50.0    1  85  3.065041
22   53.0    1  85  3.065041
23   60.0    1  85  3.065041
24   62.0    1  85  3.065041
25   65.0    1  85  3.065041
26   96.0    1  85  3.065041
27   98.0    1  85  3.065041
28  109.0    1  85  3.065041
29  112.0    1  85  3.065041
30  114.0    1  85  3.065041
31  116.0    1  85  3.065041
32  118.0    1  85  3.065041
33  120.0    1  85  3.065041
34  122.0    1  85  3.065041
35  125.0    1  85  3.065041
36  128.0    1  85  3.065041
37  128.7    1  85  3.065041
```

Data Analysis for Failure Time Data

```
38 129.3  1 85 3.065041
39 130.0  1 85 3.065041
40 133.0  1 85 3.065041
41 137.0  1 85 3.065041
42 141.0  1 85 3.065041
43 149.0  1 85 3.065041
44 152.0  1 85 3.065041
45 154.0  1 85 3.065041
46 157.0  1 85 3.065041
47 161.0  1 85 3.065041
48 172.0  1 85 3.065041
49 173.0  1 85 3.065041
50 174.0  1 85 3.065041
51 176.0  1 85 3.065041
52 176.7  1 85 3.065041
53 177.3  1 85 3.065041
54 178.0  1 85 3.065041
55 180.0  1 85 3.065041
56 182.0  1 85 3.065041
57 185.0  1 85 3.065041
58 189.0  1 85 3.065041
59 193.0  1 85 3.065041
60 197.0  1 85 3.065041
61 217.0  1 85 3.065041
62 241.0  1 85 3.065041
63 245.0  1 85 3.065041
64 249.0  1 85 3.065041
65 257.0  1 85 3.065041
66 261.0  1 85 3.065041
67 265.0  1 85 3.065041
68 297.0  1 85 3.065041
69 321.0  1 85 3.065041
70 365.0  1 85 3.065041
```

5.9 Complement: Using the Interface

In this complement, we run through the analysis of the CAF data provided in Section 5.4. Recall the data that are in the list example4a.dat0. The first five rows of the first data set have the form:

```
example4a.dat0[[1]]$mat[1:5,]
    V1 V2 V3    V4
1 168  1 85 0.980198
2 431  1 85 0.980198
3 459  1 85 0.980198
4 547  1 85 0.980198
5 563  1 85 0.980198
```

The first column is the time, and the second is an indicator variable denoting whether the time is a time for failure (1) or time taken off test without a

FIGURE 5.14
Failure time distribution GUI filled out to plot Figure 5.5.

failure yet occurring (2). The third column is temperature (identical for all experiments) and the fourth column is the function $rh/(1-rh)$ to denote the dependence of humidity (counted as a fluence here).

We first separate the data into simple matrices representing each condition, which we can then use as a basis for plots. This is done in Splus using the commands:

```
> m1<-example4a.dat0[[1]]$mat
> m2<-example4a.dat0[[2]]$mat
> m3<-example4a.dat0[[3]]$mat
```

Then we can plot the data vs. some standard failure time distributions. Left-click on the "failure time analysis" heading, and choose "single distribution analysis" from the drop-down menu. This brings up the GUI shown Figure 5.14, which we have filled out to plot Figure 5.5. Similarly, m2, would be substituted to plot Figure 5.6, and m3 to plot Figure 5.7.

The return value is a structure that will contain important information from the fit. It is shown below for this data set:

```
> ex4.49.5
$time:
 [1]  168  431  459  547  563  575  603  706  814  874  922
[12] 1040 1064 1065 1066 1160 1592 2359 2504 3176 3920

$surv:
 [1] 0.9855072 0.9710145 0.9565217 0.9420290 0.9275362
 [6] 0.9130435 0.8985507 0.8840580 0.8695652 0.8550725
[11] 0.8405797 0.8260870 0.8115942 0.7971014 0.7826087
[16] 0.7681159 0.7536232 0.7391304 0.7246377 0.7101449
[21] 0.6956522

$km:
Call: survfit(formula = Surv(ftime, cens))

  n events mean se(mean) median 0.95LCL 0.95UCL
 69     21 3208      170     NA      NA      NA
```

Data Analysis for Failure Time Data 177

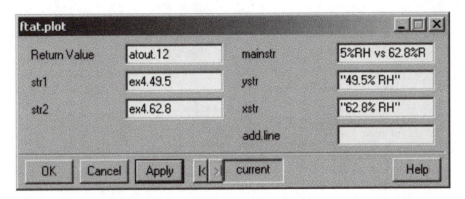

FIGURE 5.15
Acceleration transform from failure time data GUI, filled out to plot the first plot in Figure 5.8.

```
$weibull:
$weibull$par:
[1] 3099.7341842    0.7615178

$weibull$logpar:
[1] 8.039072 1.313167

$weibull$lik:
[1] -237.5728

$lognormal:
$lognormal$par:
[1] 9.261699 2.014674

$lognormal$lik:
[1] -212.6577
```

The structure, in addition to providing the Weibull and lognormal maximum likelihood estimates and likelihoods, provides the survival function in suitable form to be plotted with the acceleration transform function. Assuming the other two objects created with the failure time plotter are "ex4.62.8" and "ex4.75.4," then we can create Figure 5.8 by first formatting the graph sheet with the command:

```
>par(mfrow=c(2,2))
```

in the commands window. Then to create the first plot we left-click the "kinetic model analysis menu," then "acceleration transforms" on the first drop-down menu, and then on "acceleration transforms from survival curves." This brings up the GUI in Figure 5.15, which we have filled out to plot the acceleration transform for 49.5% relative humidity vs. 62.8% relative humidity. Note that the entries "ystr" and "xstr" are the string variables (in quotes) labeling the y and x axes, respectively.

In addition to the plot, the output provides the times plotted on the x and y axes, as well as the best linear fit through the data that runs through the origin if desired. The return value of the above is

```
> atout.12
$t1:
 [1] 3920.0000 3394.8232 3176.0000 2691.7664 2504.0000
 [6] 2397.3826 2359.0000 1783.7500 1592.0000 1261.6465
[11] 1160.0000 1086.7351 1066.0000 1065.2059 1065.0000
[16] 1064.1912 1064.0000 1044.2352 1040.0000  941.0879
[21]  922.0000  881.0588  874.0000  821.9411  814.0000
[26]  718.7060  706.0000  613.6030  603.0000  577.4707
[31]  575.0000  563.8824  563.0000  547.9412  547.0000
[36]  462.8823  459.0000  431.8236  431.0000  171.8680

$t2:
 [1] 333.4348 323.0000 321.8406 319.0000 314.5942 303.0000
 [7] 301.9565 299.0000 297.2754 292.0000 291.5362 290.0000
[13] 289.3478 287.0000 286.7971 286.0000 285.4348 283.0000
[19] 280.9131 271.0000 269.7247 263.0000 259.5218 239.0000
[25] 234.3044 203.0000 201.6087 191.0000 190.5942 187.0000
[31] 185.9565 175.0000 173.1884 150.0000 149.3043 138.0000
[37] 137.6522 130.0000 129.5362 114.0000

$coef:
         X
 0.150806
```

Here t_1 is the x-axis data, and t_2 is the data from the y axis. To fit the acceleration transform approximation 5.16, $t_1 = \phi_1(s_1, s_2)\left(1 - \exp(-\phi_2(s_2)t_2)\right)$, we construct a simple little S function. (Note that here t_1 and t_2 are the opposite of that in the data structure.)

```
> my.fit
function(phi2, str)
{
        t1 <- str$t2
        t2 <- str$t1
        x <- 1 - exp(- phi2 * t2)
        rstr <- lsfit(x, t1, int = F)
        phi1 <- rstr$coef
        ss <- sum(rstr$resid^2)
        plot(t2, t1)
        lines(t2, phi1 * x)
        list(ss = ss, phi1 = phi1, phi2 = phi2)
}
```

The simplest way to construct a Splus function is to start by writing:

```
>my.fit<-function()
```

and then using the edit function:

Data Analysis for Failure Time Data

```
>my.fit<-edit(my.fit)
```

The default edit function is a simple WYSIWYG editor. Save the edited result before quitting the editor.

Applying this function in the neighborhood of the parameters given in Section 5.3 will recreate the plots in Figure 5.9. To produce the plots in Figure 5.9, note that plotting a vector x against a vector y is done simply by typing

```
>plot(x,y)
```

in the command window. To obtain a least-squares fit of y by x, type

```
> lsfit(x,y).
```

To force the intercept to be 0, type

```
>lsfit(x,y,int=F)
```

To add the least-squares line to the plot, type in sequence

```
>plot(x,y)
>abline(lsfit(x,y,int=F))
```

The command lsfit produces a structure holding a great deal of information about the linear regression. To obtain the coefficient, simply type

```
>lsfit(x,y,int=F)$coef
```

To apply this to create Figure 5.10, type

```
> phi1vec<-c(331.98,50.14,200.44)
> phi2vec<-c(.0015,.00122,.00151)
> rhvec<-c(.495/(1-.495),.628/(1-.628),.754/(1-.754))
> rhvec1<-c(rhvec[1]/rhvec[2],rhvec[1]/rhvec[3],rhvec[2]/rhvec[3])
> plot(log(rhvec1),log(phi1vec*phi2vec))
> abline(lsfit(log(rhvec1),log(phi1vec*phi2vec),int=F))
```

To obtain the slope, type

```
> lsfit(log(rhvec1),log(phi1vec*phi2vec),int=F)$coef
```

Exercise

Recreating Figure 5.11 is left to the reader as an exercise.

To go from this point to maximum likelihood fitting of the data we need to use the freeware. First, we need to create the two one-step processes with a power law for the second variable, with independent sets of parameters,

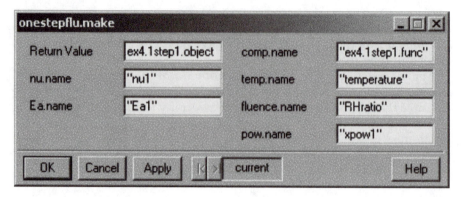

FIGURE 5.16
One-step process with second variable GUI, filled out to make one component of the reversible model.

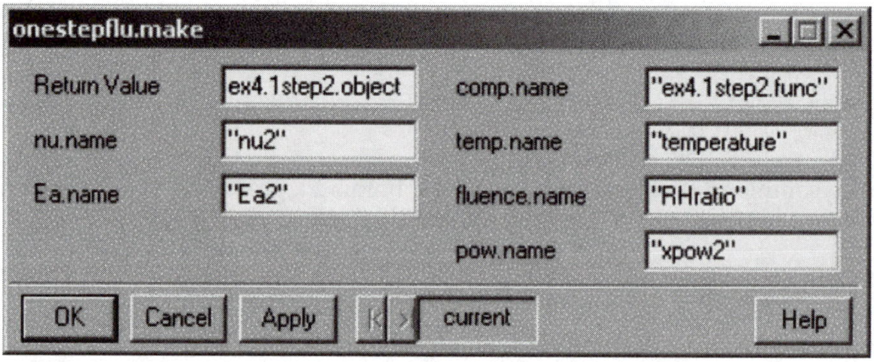

FIGURE 5.17
One-step process with second variable GUI, filled out to make another component of the reversible model.

and combine them into a reversible process. So the reader can follow the naming convention, we show the full sequence of GUIs in Figure 5.16 through Figure 5.27, with comments in the text for each and Splus commands interspersed in the correct order between the comments.

In Figure 5.16 and Figure 5.17 we show creation of the base processes for the model; in Figure 5.18 we combine them.

Now we combine the kinetic model and the data into a single object, which can be run through an automated fitting routine. Left-click the "kinetic model analysis" heading to display its drop-down menu, then left-click on "data analysis branch," and finally on "construct a failure time model data structure." This brings up the first of a sequence of GUIs, applying the first GUI once it is filled out brings up the second, and so on until four GUIs have been applied. This is necessary because the second and third GUIs adapt to the information provided before. Although it is possible to do this more

FIGURE 5.18
Reversible or equilibrium process GUI, filled out to combine the two components generated in Figure 5.16 and Figure 5.17.

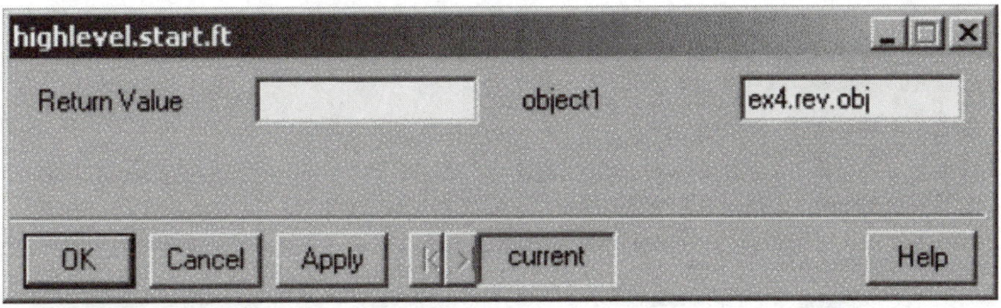

FIGURE 5.19
First GUI in series filled out to combine data with model to create a model data object to use in the fitting program; here the model is input.

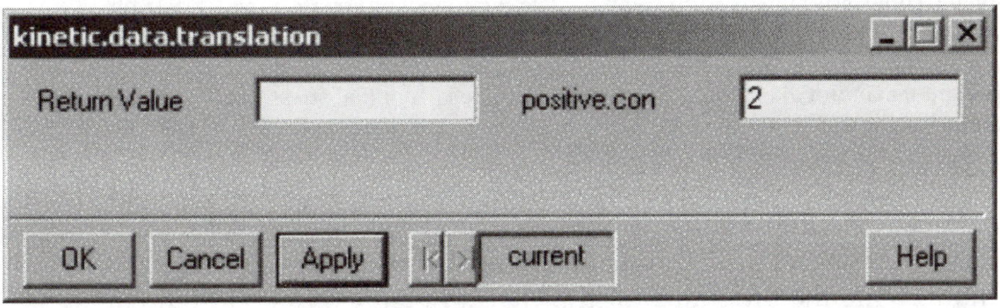

FIGURE 5.20
Second GUI in series filled out to combine data with model to create a model data object to use in the fitting program; here the states corresponding to failure or degradation are input.

elegantly using some of the extra functionality in the GUIs, the approach taken here depends primarily on the LISP-like structure in the language, rather than the specific implementation of the GUIs for different versions of Splus. Figure 5.19 through Figure 5.22 show the sequence of GUIs appropriately filled out.

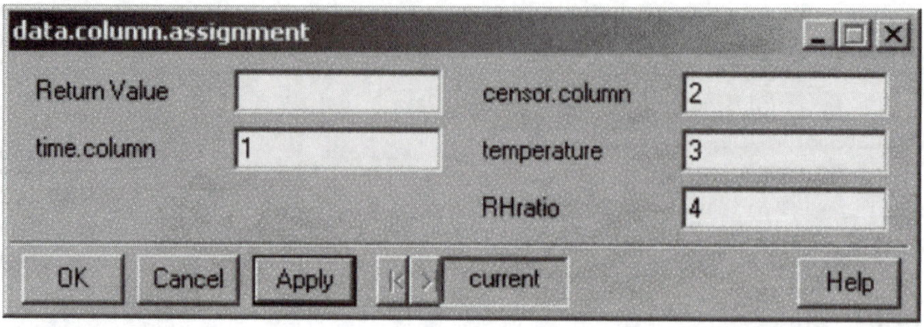

FIGURE 5.21
Third GUI in series filled out to combine data with model to create a model data object to use in the fitting program; here columns of the matrices in the data object are linked to variables in the model object.

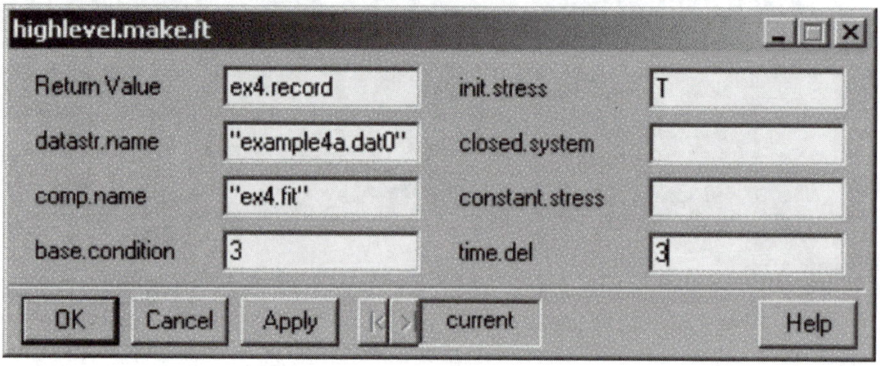

FIGURE 5.22
Final GUI in series filled out to combine data with model to create a model data object to use in the fitting program; here the object recording the programming and the data object to attach are recorded, as well as which component of the data object should be used as the base in pseudo-maximum likelihood fitting, whether there is an initial stress matrix for each data set, and what the maximum time delay might be from start of aging to start of recording.

The GUI shown in Figure 5.19 simply asks for the kinetic model object. It then uses the information in this object to print out the state vector with a rough labeling of which function created that state in the report window, and to write the programs for the GUIs in Figure 5.20 and Figure 5.21.

The report window lists:

```
[1] "ex4.1step1.func" "ex4.1step2.func":
```

The second GUI (Figure 5.20) has two slots. The important one, positive.con, simply asks what states correspond to failure-causing states.

Applying this pops up the third GUI (Figure 5.21), which takes its names from the vector of observable variables implied by the problem and the environmental variables listed in the model object. It is looking for the column numbers in the data matrices corresponding to those variables. So the failure time is column 1, and the censoring indicator is column 2 (1 is failed, 2 is censored, etc.)

The final GUI to pop up (Figure 5.22) allows all the information to be concatenated together; a "model.data.object" is created, as well as a record of what goes into it. The return value is a record of all the programs that can be used to transfer data between computers along with the Splus command "dump" and "source." The slot named "datastr.name" is the data to fit. The slot named "comp.name" is the name of the object to call to actually try fits. The slot "base.condition" tells which environmental condition should be fit to the Weibull distribution in the pseudo-maximum likelihood calculation. The slot "time.del" is the maximum time delay between start of the environmental stress and the first measurement in hours. The slot "init.stress" indicates that an initial stress matrix is included in the data structure. From some programming difficulty, we must always include an initial stress in the data structure. In this case it is a stress at which no chemical reactions will occur, e.g.,

```
> example4a.dat0[[1]]$init
     [,1] [,2] [,3] [,4]
[1,]   1    0 -200   0
[2,]   1    0 -200   0
```

shows a matrix corresponding to 2 h at −200°C 0% relative humidity. The need for this now is a bug in the programming. The 3-h delay time is a real physical parameter representing what happens in real accelerated tests. In this case, this represents the time it takes for the system to come to full equilibrium in temperature and humidity from the turn-on time. However, failure can occur before this time, so we estimate for each experimental cell what the time delay before starting should be.

Now if we type

```
>ex4.fit
```

in the commands window, this will give a long function that is used for the basis for fitting the data. Before we can check any fits, we need a starting value for the parameter vector. To obtain the full parameter vector, type

```
> ex4.record$parameter.vec
```

which returns the character vector:

```
[1] "nu1"       "Ea1"        "xpow1"       "nu2"
[5] "Ea2"       "xpow2"      "time.unc1"   "time.unc2"
[9] "time.unc3" "weib.alpha" "weib.beta"
```

There are six physical parameters, three nuisance parameters corresponding to the uncertain starting times, and the scale and shape parameters for

FIGURE 5.23
Failure time analysis GUI filled out, in this case to obtain the Weibull parameters for pseudo-maximum likelihood.

the Weibull distribution. The literature provides us with an estimate of the two activation energies, the values of nu1, nu2 and each power we can obtain from our preliminary estimates developed in Section 5.3. Each of the time uncertainty parameters enters into the equations in the form:

```
tdel = 3*exp(-time.uncI)
```

So we can simply start with 1 (or 100 to ignore the effect) to begin.

We can obtain the Weibull parameters from the failure time analysis function used in the beginning to look at the data. Recall m3 is the data from the third condition (if this analysis exercise is followed from the beginning). To obtain the Weibull parameters we left-click on the "failure time analysis" heading, and choose "single distribution analysis" from the drop-down menu. Then we fill out the "ft.plot" GUI to fit m3 as shown in Figure 5.23.

Typing "dum$weibull" then gives us

```
> dum$weibull
$par:
[1] 140.423596   1.541225
$logpar:
[1] 4.9446635 0.6488345
$lik:
[1] -400.362
```

With this we can fill in a starting parameter vector as

```
> ex4.parvec00<-c(4.93,0.9,2.2,-6.44,0,0,100,100,100,140.42,1.54)
```

To obtain a plot of the fit with this vector and a measure of the fit, we first reset the graph sheet to plot only one plot on the page with the command:

```
>par(mfrow=c(1,1))
```

Then we can obtain a plot and the negative of the likelihood by typing

Data Analysis for Failure Time Data

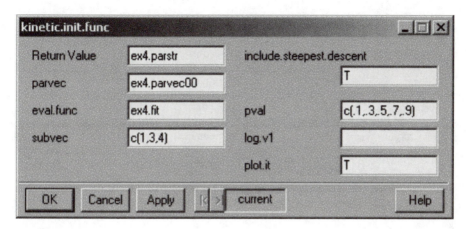

FIGURE 5.24
One-step Gauss–Newton optimization GUI filled out to work from our starting vector.

```
> ex4.fit(ex4.parvec00,T)$ss
[1] 1061.527
```

This produces the plot on the left of Figure 5.12.

To optimize, there are two options: there is a program that calculates one step in a modified Gauss–Newton fit. This is accessed under the "kinetic model analysis" drop-down menu, on the "data analysis branch" at "do one steepest descent trial." The GUI for the "kinetic.init.func" is filled out in Figure 5.24 to do one Gauss–Newton step. Alternatively, left-clicking "fit the model" brings up the GUI for the function "powell.kinetic.fit." This function runs through as many iterations of a conjugate gradient optimization as we choose to specify. We can also choose to have the first direction picked by the one-step Gauss–Newton function. This GUI filled out for three full iterations is shown in Figure 5.25. In both search functions the slot "subvec" specifies the components of the parameter vector the optimization search should run through. In Figure 5.24, the slot "pval" gives the proportions of the Gauss–Newton step to check through. The slot "log.v1" contains default values of diagonal biases to add to the matrix guiding the steepest descent. The slot "plot.it" if T causes the function to plot the fit if it is improved.

The powell.kinetic.fit function shown in Figure 5.25 begins by trying one full search through the parameters (which takes n^2 iterations for n parameters. Then it walks through until the first of "nsrch"* "num.rep" total attempts or "nsrch" attempts without improvement. Specifying "steep.desc.init" forces the first direction of the conjugate gradient search to be the best Gauss–Newton direction on each full search. The parameter "sigma" specifies the standard deviation of the random-sized step to take in attempting to optimize along the conjugate gradient. Typically, this number should be between 0.001 and 0.03.

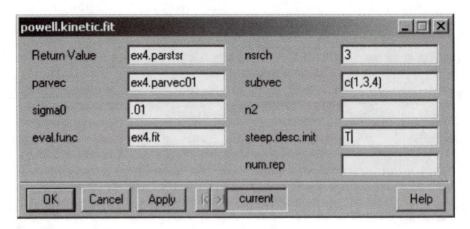

FIGURE 5.25
Conjugate gradient optimization GUI filled out to work from an improved parameter vector.

To go through only one conjugate gradient search, we can put −1 in nsrch. To do only part of the conjugate gradient search, specify in n2 how many steps you wish to take (then n2*n steps will occur). The algorithm used is spelled out in Powell (1964).

This program is the main driver in this system for optimization. However, it is also cranky. The parameters will necessarily stay the same sign during the search that they start out, but they can proceed to nonphysical and even noncalculable parameter values. The program can also take a long time to converge, so it may be useful to set up the freeware on multiple computers, and work on one while the other is running. (This difficulty and the intermediate friendliness of the interface are the reasons the software is free.)

To aid in transferring a problem from one computer to another, there is a command on the "kinetic model analysis" drop-down menu to "save the model." This brings up the GUI for the function "workspace.file," which allows us to save the model and any parameter values we have fit. The GUI is shown in Figure 5.26 filled out to save the model plus the parameter vectors "ex4.parvec00" and "ex4.parvec01."

After each full run of an optimization routine, the best parameter vector found by the program comes out as a component of the output, so, for example, to save a new value from an optimization run, type

```
> ex4.parvec02<-ex4.parstr$parvec
```

and the new parameter vector is saved in ex4.parvec02, which can then be used in another round of optimization.

The reason to have the one-step Gauss–Newton optimizer as well as the conjugate gradient optimizer is that sometimes the Gauss–Newton convergence is very fast, so it is actually better to go through a few steps using the Gauss–Newton optimizer before running the other optimizer to see if we are done.

Data Analysis for Failure Time Data

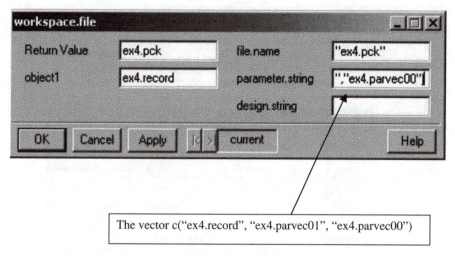

FIGURE 5.26
Model-saving GUI, saving programming and analysis record for transfer to other Splus environments loaded with this freeware.

The pseudo-maximum likelihood estimate is very close to

```
> ex4.parvec01
 [1]     5.461496     0.900000     2.378368    -6.621840
 [5]     0.000000     0.000000   100.000000   100.000000
 [9]   100.000000   140.420000     1.540000
```

which gives the fit shown on the right in Figure 5.12.

Once the data are fit, we can now make predictions and calculate confidence intervals for those predictions. The GUI to make those predictions is accessed in the "kinetic model analysis" menu on the "data analysis branch" at "fast parameter covariance estimates, failure time models." To do the prediction, we will need a matrix that asks for a 20-year prediction, but with emphasis on the early time to see the action in this function. The prediction matrix has nearly the same structure as the data matrix used in the fit structure; only the time column has increments of time rather than cumulative time. For example,

```
> ex4.predmat<-cbind(c(1,10,100,1000,8760-
1111,4*8769,15*8760),rep(0,7),rep(40,7),rep(1,7))
```

creates the matrix:

```
> ex4.pred.mat
      [,1] [,2] [,3] [,4]
[1,]     1    0   40    1
[2,]    10    0   40    1
[3,]   100    0   40    1
```

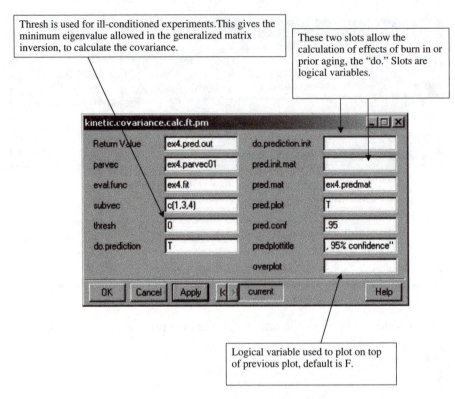

FIGURE 5.27
Predictions and confidence bound creating GUI, filled out to use the optimized model to look at 40°C, 50% relative humidity, 20 years.

```
[4,]   1000   0  40  1
[5,]   7649   0  40  1
[6,]  35040   0  40  1
[7,] 131400   0  40  1
```

Summing the total in the time column gives 20 years of hours. In Figure 5.27 we have filled in the GUI to do the prediction at 95% confidence (Figure 5.13) and produce the data structure ex4.pred.out.

```
> ex4.pred.out
$cov:
            [,1]           [,2]           [,3]
[1,]  0.0010833062 -0.0002945596  0.0005036527
[2,] -0.0002945596  0.0039715274 -0.0010245511
[3,]  0.0005036527 -0.0010245511  0.0043656825

$prediction.covariance:
     [,1]         [,2]          [,3]          [,4]
[1,]  0  0.000000e+000  0.000000e+000  0.000000e+000
[2,]  0  7.254697e-013  2.946638e-013  2.127148e-011
[3,]  0  2.946638e-013  1.643761e-012  5.700270e-011
```

Data Analysis for Failure Time Data

```
[4,]    0 2.127148e-011 5.700270e-011 2.158377e-009
[5,]    0 3.190163e-011 1.741690e-010 6.051038e-009
[6,]    0 1.329096e-011 1.804377e-010 5.944147e-009
[7,]    0 2.481054e-011 1.798253e-010 6.114004e-009

              [,5]          [,6]           [,7]
[1,] 0.000000e+000 0.000000e+000 0.000000e+000
[2,] 3.190163e-011 1.329096e-011 2.481054e-011
[3,] 1.741690e-010 1.804377e-010 1.798253e-010
[4,] 6.051038e-009 5.944147e-009 6.114004e-009
[5,] 1.845521e-008 1.909948e-008 1.904635e-008
[6,] 1.909948e-008 2.034664e-008 1.994998e-008
[7,] 1.904635e-008 1.994998e-008 1.975461e-008

$prediction:
[1] 0.000000e+000 2.267672e-008 6.536219e-007 2.027458e-005
[5] 3.387029e-004 5.949774e-004 5.945072e-004 5.948643e-004

$low.bound:
[1]  0.000000e+000  2.267672e-008 -7.473741e-007  1.816572e-005
[5]  2.622857e-004  3.715241e-004  3.598826e-004  3.636783e-004

$upper.bound:
[1] 0.000000e+000 2.267672e-008 2.054618e-006 2.238343e-005
[5] 4.151201e-004 8.184307e-004 8.291318e-004 8.260502e-004

$pred.time:
[1]      0      1     11    111   1111   8760  43800 175200

$wb.contribution:
             [,1]           [,2]         [,3]
[1,] 0.00008046803 0.00001832905 0.0002291450
[2,] 0.00001832905 0.00076842719 0.0008046578
[3,] 0.00022914503 0.00080465778 0.0013933812
```

The data structure has seven components:

1. $cov: The estimated pseudo-maximum likelihood covariance for the free kinetic parameters
2. $prediction.covariance: The estimated covariance matrix for the predictions
3. $prediction: The actual predicted probabilities of failure at each time
4. $low.bound: The lower bound on the predictions at the specified confidence level
5. $upper.bound: The upper bound on the predictions at the specified confidence level
6. $pred.time: The time corresponding to the predictions
7. $wb.contribution: The contribution of the uncertainty in the Weibull parameters to the total covariance

5.10 Complement: Exercises to Explore Some Questions in Experiment Design

The theory of experiment design presented in Chapters 2 and 4 is particular to the problem of extrapolating a real phenomenon (observed failure or degradation, or lack of either) from an easily accessible condition (accelerated test conditions) to more economically important but less accessible conditions (operating conditions for operating life). It was assumed that whatever was observed at accelerated conditions was modeled with very little statistical error and parsimoniously (using the fewest number of parameters possible) and predicts no observable degradation or failure at operating conditions.

In the final complement of this chapter and in the next chapter, we consider some complementary problems:

1. We have fit a model, but the uncertainty in the estimates or in the predictions is too large. What is the most economical way to improve the experiments?
2. There are alternative physical models (not evanescent process models, but models that provide completely different explanations and extrapolations of the observed phenomena). It is not possible to effectively distinguish between the two (or more) models given the data. What experiments will improve our ability to distinguish between the models?

A good explanation of the fundamental theory of experiment design for problem 1 can be found in Chernoff (1972). Nelson (1990) and Meeker and Escobar (1998) provide a very nice overview of very sophisticated design theory for the case when the models involve a single rate constant, i.e., when an "acceleration factor" is the appropriate summary for the way time transformations occur between different stress conditions. Problem 2 can also be approached by the methods of Chernoff (1972), once we have some idea of what conditions to check.

In this complement we take a very simple approach. We use physical reasoning to try to identify appropriate conditions for better estimates/better differentiation between models, and then use simulation to check how well they do. By combining the inherent capabilities of Splus with the tools in the freeware, it is possible and fairly convenient to simulate alternative experiments that are consistent with given models fitting the available data, to add the simulations to the data, and to see how the discrimination ability or the errors change with the new data.

Three problems are presented below. Following, a series of suggestions are provided based on the inherent structure in Splus and the data structures used in the freeware. However, the application is left to the reader.

Data Analysis for Failure Time Data

Problem 5.10.1

The pseudo-maximum likelihood approach separates the data into two sections, one used to estimate the (Weibull, in our case) distribution at the base condition and one used to estimate the kinetic parameters (the data at the nonbase condition). If we had 100 additional experimental units to put at relative humidities between 75% and 50%, with temperature and voltage identical to the original experimental conditions, where should we place them to reduce the prediction variance optimally at 35°C, 40% relative humidity?

Problem 5.10.2

Early in the study of this phenomenon, there was an alternative physical model giving rise to exactly the same acceleration transform under constant stress conditions (model 2, competing reactions, in Section 5.2). This model was the competing process model with the same model for the two rate constants. If we could do one experiment with 100 samples to help distinguish between the two models, with relative humidity between 49.5 and 75%, what would it be? We have only 5000 h for the experiment. (*Hint*: Consider changing the stress in mid experiment. What are the consequences for the two different models?)

Problem 5.10.3

Think of building an evanescent process map around the current model (using the same form for the rate constants). Assume 0.01 probability of failure or less is good enough to define a risk orthogonal experiment. Find at least three experiments running at least 5000 h, which are risk orthogonal to the competing process model above. Find three experiments running at least 5000 h, which are risk orthogonal to the reversible process model. Is it possible to think of an alternative approach to defining risk orthogonal that would not restrict experiments as much?

Hints for Problem 5.10.1

First there is a rule of thumb that should be applied that follows from the kind of analysis Chernoff (1972) originally did on the problem, which arose from some analysis by Elfving (1952). The rule of thumb is that when extrapolating, always put more experimental units closer to the place we are extrapolating to. Second, we can do some very simple analysis based on the result of Equation 5.44 and the output of the prediction program. Recall that Equation 5.44 gives the covariance for the parameters as

$$\Sigma = I_{11}^{-1} + I_{11}^{-1} I_{12}' \Sigma(\tilde{\phi}_2) I_{12} I_{11}^{-1}$$

The first term on the right is the covariance matrix for the kinetic parameters if we know the Weibull distribution. The second is the contribution of the uncertainty in the Weibull parameters to the uncertainty in the kinetic parameters. The output of the prediction program gives both the right-side term and the second term on the left side. Below we look at the decomposition:

```
> ex4.pred.out$cov
             [,1]           [,2]          [,3]
[1,]  0.0010833062 -0.0002945596  0.0005036527
[2,] -0.0002945596  0.0039715274 -0.0010245511
[3,]  0.0005036527 -0.0010245511  0.0043656825
> ex4.pred.out$wb.contribution
             [,1]           [,2]          [,3]
[1,] 0.00008046803 0.00001832905 0.0002291450
[2,] 0.00001832905 0.00076842719 0.0008046578
[3,] 0.00022914503 0.00080465778 0.0013933812
> ex4.pred.out$cov-ex4.pred.out$wb.contribution
             [,1]           [,2]          [,3]
[1,]  0.0010028381 -0.0003128886  0.0002745077
[2,] -0.0003128886  0.0032031002 -0.0018292089
[3,]  0.0002745077 -0.0018292089  0.0029723012
```

The contribution of the Weibull estimates provides at most a third of the variation. Because the Weibull estimates are estimated at the highest stress condition, farthest from operating conditions, both this result and the rule of thumb point to experiments with more data at the two lower conditions.

To simulate data at a new stress condition under this model we run a prediction, then use the *probability integral transform* to generate data consistent with the model. In particular, suppose we wish to simulate data at 60% relative humidity. Create the simulation matrix:

```
ex4.sim.mat<-cbind(c(rep(1,10),rep(10,9),rep(100,9),rep(1000,9)),
rep(0,37),rep(85,37),rep(.6/.4,37))
```

This gives

```
> ex4.sim.mat
numeric matrix: 37 rows, 4 columns.
       [,1] [,2] [,3] [,4]
 [1,]    1    0   85  1.5
 [2,]    1    0   85  1.5
 [3,]    1    0   85  1.5
 [4,]    1    0   85  1.5
 [5,]    1    0   85  1.5
 [6,]    1    0   85  1.5
 [7,]    1    0   85  1.5
 [8,]    1    0   85  1.5
 [9,]    1    0   85  1.5
[10,]    1    0   85  1.5
[11,]   10    0   85  1.5
```

Data Analysis for Failure Time Data

```
[12,]    10    0   85   1.5
[13,]    10    0   85   1.5
[14,]    10    0   85   1.5
[15,]    10    0   85   1.5
[16,]    10    0   85   1.5
[17,]    10    0   85   1.5
[18,]    10    0   85   1.5
[19,]    10    0   85   1.5
[20,]   100    0   85   1.5
[21,]   100    0   85   1.5
[22,]   100    0   85   1.5
[23,]   100    0   85   1.5
[24,]   100    0   85   1.5
[25,]   100    0   85   1.5
[26,]   100    0   85   1.5
[27,]   100    0   85   1.5
[28,]   100    0   85   1.5
[29,]  1000    0   85   1.5
[30,]  1000    0   85   1.5
[31,]  1000    0   85   1.5
       [,1] [,2] [,3] [,4]
[32,]  1000    0   85   1.5
[33,]  1000    0   85   1.5
[34,]  1000    0   85   1.5
[35,]  1000    0   85   1.5
[36,]  1000    0   85   1.5
[37,]  1000    0   85   1.5
```

Use this in the prediction GUI (Figure 5.27). This gives us a set of probabilities and times for the output, up to 10,000 h. To simulate data, we draw a set of random variables from the uniform distribution in Splus, using the native commands, and then apply the Splus approximation function. The relevant output of the prediction function is shown below, followed by the syntax for the random sampling and approximation function commands.

```
> ex4.sim.out$prediction
 [1] 0.00000000000 0.00004083609 0.00011880988 0.00022199160
 [5] 0.00034522863 0.00048659584 0.00064407489 0.00081580937
 [9] 0.00100149341 0.00119961415 0.00141017726 0.00406773307
[13] 0.00753196830 0.01162937579 0.01625046050 0.02132030990
[17] 0.02677603589 0.03257291473 0.03866751864 0.04502406050
[21] 0.11756810233 0.19560730886 0.27059503887 0.33877099102
[25] 0.39881427945 0.45068052497 0.49495736345 0.53247477944
[29] 0.56413179811 0.70289953149 0.72927894266 0.73463418436
[33] 0.73574062970 0.73597154232 0.73602070075 0.73602854977
[37] 0.73603226759 0.73603144141
> ex4.sim.out$pred.time
 [1]    0     1     2     3     4     5     6     7     8
[10]    9    10    20    30    40    50    60    70    80
[19]   90   100   200   300   400   500   600   700   800
[28]  900  1000  2000  3000  4000  5000  6000  7000  8000
[37] 9000 10000
```

First we note that the predicted probability has decreased between 9000 and 10,000 h. This points out a round-off error in the software. Because the probability of incremental failure after 9000 h is pretty much 0, we truncate both data sets early at 9000 h in the approximation. The commands to do this are as follows:

```
> ex4.sim.out$prediction<-ex4.sim.out$prediction[1:37]
> ex4.sim.out$pred.time<-ex4.sim.out$pred.time[1:37]
```

The commands below produce a simulation of 100 data points sampled from this distribution:

```
> ptime<-ex4.sim.out$pred.time
> pred60<-ex4.sim.out$prediction
> approx(pred60,ptime,sort(runif(100)),rule=2)$y
  [1]    22.78641    25.62934    45.08501    55.39705    58.99345
  [6]    75.08201    78.06612    79.04946   102.24001   102.34408
 [11]   105.56363   169.47223   173.42758   212.68976   213.18930
 [16]   224.80986   244.05597   281.27008   335.38971   346.93115
 [21]   347.59967   359.40549   372.60806   375.44180   376.87036
 [26]   382.86798   411.91342   429.26501   434.60083   436.38080
 [31]   441.54205   451.53183   462.83789   466.91666   477.27322
 [36]   482.39706   487.04205   489.02573   492.56039   492.94763
 [41]   519.32184   554.38904   624.07794   669.00000   687.41229
 [46]   697.98755   734.54346   774.62866   814.24139   885.59747
 [51]   893.74646   933.63550   958.66620   970.17773   973.69147
 [56]  1021.60870  1056.66589  1063.48816  1067.63818  1080.57568
 [61]  1106.38855  1154.18433  1173.44092  1259.32373  1269.47864
 [66]  1319.59216  1466.39636  1571.98462  1587.48975  1603.77783
 [71]  1680.19348  1749.00391  1860.39771  2377.35913  2898.34888
 [76]  9000.00000  9000.00000  9000.00000  9000.00000  9000.00000
 [81]  9000.00000  9000.00000  9000.00000  9000.00000  9000.00000
 [86]  9000.00000  9000.00000  9000.00000  9000.00000  9000.00000
 [91]  9000.00000  9000.00000  9000.00000  9000.00000  9000.00000
 [96]  9000.00000  9000.00000  9000.00000  9000.00000  9000.00000
```

The syntax is the first term in the approx command is x, the second term y, the third term is the new x variables that we want the y values for. The command rule = 2 says that when the new x variables are beyond the extremes of x, put the appropriate extreme in. Thus, approximately 25% of the data is beyond the 9000 point, consistent with the way the probability truncates. To create a data matrix we can use

```
> y1<-approx(pred60,ptime,sort(runif(100)),rule=2)$y
> y1
 [1]    8.390953    25.285160    63.929420    68.647408
 [5]   88.771248   113.691467   148.035675   150.882538
 [9]  152.701920   158.587494   162.707520   167.355988
[13]  194.353760   207.058960   209.636444   233.187119
[17]  266.802582   307.420593   328.855225   329.365265
[21]  333.180450   347.534393   354.587463   364.550690
```

Data Analysis for Failure Time Data

```
[25]  366.097198  366.592804  370.734924  372.836334
[29]  412.497131  458.503845  470.383881  479.671783
[33]  534.826721  542.768555  547.200439  547.519775
[37]  547.906006  562.673706  582.111572  584.129578
[41]  610.036926  654.472473  674.229126  687.807068
[45]  713.220459  715.812805  722.168335  743.081848
[49]  757.378479  799.557495  825.324463  851.959656
[53]  859.801575  864.176880  901.638245  950.887451
[57]  989.556335 1065.795532 1095.381470 1140.053711
[61] 1151.089966 1176.318359 1183.384033 1204.319214
[65] 1256.802490 1271.955322 1305.368774 1354.435913
[69] 1676.147339 1753.614868 1791.846802 1967.591675
[73] 1972.472412 2186.195313 9000.000000 9000.000000
[77] 9000.000000 9000.000000 9000.000000 9000.000000
[81] 9000.000000 9000.000000 9000.000000 9000.000000
[85] 9000.000000 9000.000000 9000.000000 9000.000000
[89] 9000.000000 9000.000000 9000.000000 9000.000000
[93] 9000.000000 9000.000000 9000.000000 9000.000000
[97] 9000.000000 9000.000000 9000.000000 9000.000000
> new.dat<-cbind(y1,c(rep(1,75),rep(2,25)),85,.6/.4)
```

This creates a data matrix; the first 75 times are failure times, and the last 25 times are censored times.

To create a new data set, which includes this new data matrix, we create a copy of example4a.dat0, and add to it in a way consistent with the data structure.

```
> example4b.dat0<-example4a.dat0
> length(example4b.dat0)
[1] 3
> attributes(example4b.dat0[[1]])
$names:
[1] "mat"          "init.stress"

> exdat.str<-example4b.dat0[[1]]
> exdat.str$mat<-new.dat
> example4b.dat0[[4]]<-exdat.str
```

This series of commands adds the new data to the original data set, creating a new data set that can be fit, and for which the prediction covariance can be measured. At the same time, the format of the input data structure is maintained.

From this point, the reader should be able to choose and simulate any combination of additional experimental points and compare the output covariance structures. To extract the diagonal elements of the covariance matrix and to compare to prediction variance, the command

```
> diag(ex4.sim.out$prediction.covariance)
[1] 0.000000e+000 0.000000e+000 1.530047e-010 3.891639e-010
[5] 5.861159e-010 2.194329e-009 3.790476e-009 4.377008e-009
[9] 5.977555e-009 1.135698e-008 7.660670e-008 2.888526e-007
```

```
[13]  6.825755e-007  1.178039e-006  2.162297e-006  3.150631e-006
[17]  4.769618e-006  6.665384e-006  8.943232e-006  5.676467e-005
[21]  1.545131e-004  2.908350e-004  4.611621e-004  6.586149e-004
[25]  8.777145e-004  1.116638e-003  1.354771e-003  1.594620e-003
[29]  3.445721e-003  4.264737e-003  4.550252e-003  4.632490e-003
[33]  4.656474e-003  4.664574e-003  4.662343e-003  4.660862e-003
[37]  4.666429e-003
```

extracts just the diagonal elements (the variances) from the covariance matrix. Taking a ratio of the variances from the two experiments gives a comparison of the relative efficiency of the two experiments. In particular:

$$\mathit{eff}\left(\frac{\text{expt 2}}{\text{expt 1}}\right) = \frac{\text{var}(\text{expt1})}{\text{var}(\text{expt 2})}$$

This definition is used because the variance varies inversely with the sample size, so the efficiency larger than 1 means that experiment 2 requires fewer samples to achieve the same variance as experiment 1.

Hint for Problem 5.10.2

Construct the fit for the competing process model. Consider simulating experiments for both models in which the humidity is either increased or decreased in a step after a certain time. Use the overplot option on the prediction GUI to plot the predictions for the two models on the same experiment. When the confidence bounds clearly separate, that is an experiment that can tell the two models apart.

Hints for Problem 5.10.3

No hints should be required.

6

Data Analysis for Degradation Data

Some material systems show no easily observable signs of internal degradation phenomena until the systems fail. The Conductive Anodic Filament (CAF) failure mode that was the subject of the data analysis in the previous chapter is an example. Other material systems show continuous degradation over time. In this chapter, we examine how kinetic models can be used in the analysis of such data, and what sort of models we need of how the kinetics translates to the observable degradation. We also analyze an example data set.

6.1 Motivation and Models

With the advent of computerized data acquisition systems, the amount of data that could be collected on individual units undergoing tests over time increased by orders of magnitude. Quite often, degradation data consist of a series of curves, surfaces, or hypersurfaces measured at different time intervals. Figure 6.1 shows an artificial example of two "spectra" measured on a device at different times.

Assume the upper curve is the later curve. In studying this kind of change, the first question is how to summarize it. Clearly, the bump centered near 520 wavelength units is not changing, while that near 560 is growing. But is the growth at 540 units proportional to the growth at 560? Is there a way of normalizing so we would expect it to be? What do we analyze over time, an integrated difference, or do we simultaneously look at changes at several different points?

The answer to this sort of question is often as much an issue of the physics governing the phenomena as it is of the statistics. Thus, sometimes it is necessary to guess, or to start with a potentially very high information version, and see if in fact all the information points to a single phenomenon. In the example of studying the spectra above, a high information version would be to study an appropriate normalization of the difference (or ratio, again, depending on the physics) at a number of different wavelengths. If all the changes obey the same law it should show up naturally in the analysis.

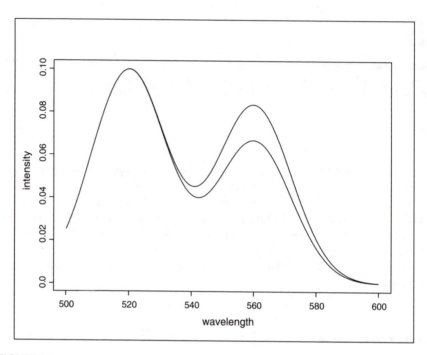

FIGURE 6.1
Artificial example of raw degradation data; a whole spectrum changes over time.

The basic model we would like to use for degradation data takes the following form:

$$\frac{d}{dt} A_{t\, m\times 1} = K(s)_{m\times m}\, A_{t\, m\times 1}$$
$$Y_{r\times 1} = f\left(M_{r\times m}\, A_{t\, m\times 1}\right) + \varepsilon_t \tag{6.1}$$

As throughout the book A_t is an $m \times 1$ state vector, $K(s)$ is an $m \times m$ vector of functions of stress, Y is an observable, this time an $r \times 1$ vector, and ε_t represents a time series of noise vectors. The new terms f and M provide a mathematical model of how the internal state is translated to an observable.

At this level of generality it is extremely important to be able to identify whether a phenomenon is caused by a difference of initial state vector in the device, A_0, or by the structure of $K(s)$, the structure of the rate constants in the differential equations and their dependence on stress. As in the rest of the book, we are confining the modeling to first-order kinetics. With the full level of generality, a number of devices are required at each stress condition, so the full range of variability caused by device to device variation can be characterized. In particular, for unknown (monotone) f, acceleration transforms need to be estimated both between devices with different initial state vectors and between different stress conditions for devices with similar initial state vectors. If the devices are manufactured in batches, it is important that devices from the same batch be distributed across many different experimental cells, because the batch manufacturing often leads to more uniformity within the batches.

One way to identify devices with similar initial conditions is to measure their degradation through a series of step stress experiments, each device following an identical trajectory. Devices with similar degradation paths will have similar initial conditions. Stepping similar devices apart will then allow estimation of the acceleration transforms appropriate to $K(s)$.

Unfortunately, the data we have seen to date actually conform to a simpler model:

$$\frac{d}{dt} A_{t\,m\times 1} = K(s)_{m\times m} A_{t\,m\times 1} \qquad (6.2)$$

$$Y_{r\times 1} = M_{r\times m} A_{t\,m\times 1} + \varepsilon_t$$

This is unfortunate because we have no examples of the fully general model. However, it does lead to simpler treatment, which is certainly a plus.

Let ω correspond to a particular individual device and $\rho(\omega)$ correspond to the distribution of initial conditions between devices. Then an individual degradation path following the stress trajectory

$$\begin{pmatrix} s_1 & s_2 & & s_n \\ t_1 & t_2 & \cdots & t_n \end{pmatrix}$$

will obey the law:

$$Y_t(\omega) = M \prod_{s=s_1}^{s_n} \exp(-K(s_i)t_i) A_0(\omega) \qquad (6.3)$$

And the average degradation path obeys the law:

$$\int Y_t(\omega) d\rho(\omega) = \int M \prod_{s=s_1}^{s_n} \exp(-K(s_i)t_i) A_0(\omega) d\rho(\omega)$$

$$= M \prod_{s=s_1}^{s_n} \exp(-K(s_i)t_i) \int A_0(\omega) d\rho(\omega) \qquad (6.4)$$

Thus, the kinetics can be investigated by modeling the average degradation paths, at each condition. The freeware we have developed and the statistical theory (Complement 6.4) focuses on this case.

Exercise 6.1.1

Use the theory in Complement 6.4.1 to show how to use Equation 6.3, at different times, and the estimates of M and K to estimate $A_0(\omega)$ for each device.

6.2 Background for the Example

In this example, we model the change in loss associated with the exposure of a specially doped silica optical fiber to ionizing radiation. For proprietary reasons, all data have been rescaled, and some have been subjected to other transformations.

Silica optical fiber is an amorphous form of silica dioxide with a number of dopants added to change various of its optical properties. The effect of high-energy radiation (Griscom et al., 1991; Erdogan et al, 1994) on silica is to create sets of metastable defects, which anneal at different rates. Experience with ultraviolet-induced gratings is that the annealing is governed by a distribution of activation energies (Erdogan et al., 1994; LuValle et al., 1998), with a single premultiplying constant. The defects are absorbing, so loss along the fiber is governed by a differential equation:

$$\frac{dP}{dz} = -Pk[defect]_z \tag{6.5}$$

where $[defect]_z$ is the concentration at point z, and z is the direction along the fiber. Assuming constant defect concentration along the fiber, this translates to

$$P_z = P_0 \exp(-k[defect]z) \tag{6.6}$$

where P_z is the power out at length z. Because loss is measured in decibels per unit length where decibels are defined by

$$\alpha_{dB} = 10\log_{10}\left(\frac{P_z}{P_0}\right) \tag{6.7}$$

We see that loss measured in decibels is linearly proportional to the concentration of defects. Thus, we see that not only is Equation 6.3 appropriate for the loss, but also $A_0(\omega)$ is especially simple; everything starts out in the defect-free state.

TABLE 6.1

Experimental Conditions Producing the Data

Experiment	Exposure Time	Exposure Dose Rate	Exposure Temperature	Anneal Temperature
1	8.333	0.6	19	110
2	8.333	0.6	19	160
3	1.173	0.639	19	110
4	1.173	0.639	19	160
5	160	0.005	19	110
6	160	0.005	19	160
7	320	0.0025	19	110
8	320	0.0025	19	160

TABLE 6.2

Measurements at a Dose Rate of 5e–4 Units

Time (hours)	Loss
500	0.0181
1000	0.0340
1500	0.0475

Properly, the loss at each wavelength for each kind of defect obeys the laws given in Equations 6.5 through 6.7, so the data are squarely in the realm of model 6.2 as long as the kinetics is first order, or can be so approximated (Zwolinsky and Erying, 1947).

Experimentally, we exposed fiber sections to varying dose rates and doses, and then annealed them. Ideally, we would have recorded optical spectra continuously through both the irradiating and annealing processes. For practical considerations, we have only the dose rate and dose information and measurements during the annealing. The experiment design is shown in Table 6.1.

We assume that the uncertainty in starting time of the anneal can be up to 6 min based on handling times and some of the data transformations. In addition, for model confirmation we have loss measured at a dose rate of 0.0005 at 25°C for 500, 1000, and 1500 h given in Table 6.2 (also adjusted for consistency).

6.3 Data Analysis for the Example

The data are stored in "example3a.dat0." The first few rows of the data matrix for the first data set are as follows:

```
> example3a.dat0[[1]]$mat[1:3,]
          [,1]       [,2]       [,3]       [,4]       [,5]
param 0.00000000 0.00000000 0.00000000 0.00000000 0.00000000
param 0.04333333 0.06794357 0.07403219 0.07943559 0.08451605
param 0.08638889 0.09326613 0.10342741 0.11798394 0.13096786

[,6]  [,7] [,8] [,9]
param 110.1  24   0    1
param 109.9  24   0    1
param 110.0  24   0    1
```

The first column is elapsed time, the second through fifth are the incremental change in losses at different wavelengths, the sixth is the temperature of the fiber, the seventh is the temperature of the measurement instrument, the eighth is the dose rate, and the ninth is a statistical weight corresponding to the precision of the measurement. The initial stress matrix has the form:

```
> example3a.dat0[[1]]$init
       [,1] [,2] [,3] [,4] [,5] [,6] [,7] [,8]
[1,] 4.1665   0    0    0    0   19    0   0.6
[2,] 4.1665   0    0    0    0   19    0   0.6
```

using the same format, except that column 9 is not included.

The basic physical model we assume is a single family of defects (with a distribution of activation energies for annealing) that are generated, possibly by a power law of dose rate. We note that a power law in dose rate may mean that there is some complicated nonlinear kinetics going on in the generation of the defects, but using the power law allows us to approximate the rate-limiting kinetics by a first-order kinetic process. The specific physical model translates to a mathematical model of the form:

$$A_1 \underset{k_{2i}}{\overset{k_1}{\rightleftarrows}} \begin{pmatrix} A_2 \\ \cdot \\ \cdot \\ \cdot \\ A_n \end{pmatrix} \text{ or alternatively}$$

$$\frac{d}{dt}\begin{pmatrix} A_1 \\ \cdot \\ \cdot \\ \cdot \\ A_n \end{pmatrix} = \begin{pmatrix} -k_1 & k_{21} & \cdot & \cdot & k_{2n} \\ k_1 p_2 & -k_{21} & \cdot & \cdot & 0 \\ \cdot & 0 & \cdot & \cdot & \cdot \\ \cdot & \cdot & \cdot & \cdot & 0 \\ k_1 p_n & 0 & \cdot & \cdot & -k_{2n} \end{pmatrix}\begin{pmatrix} A_1 \\ \cdot \\ \cdot \\ \cdot \\ A_n \end{pmatrix} \tag{6.8}$$

with

$$k_1 = (\text{dose rate})^{ypow}(k_1)$$

$$k_{2i} = v \exp\left(-\frac{E_{ai}}{kT}\right) \tag{6.9}$$

Data Analysis for Degradation Data

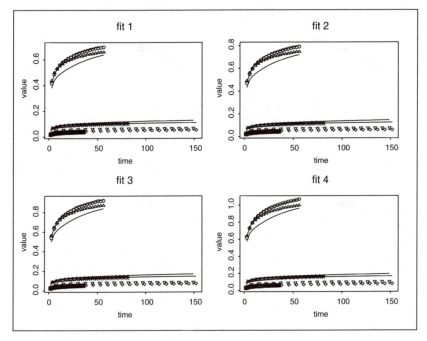

FIGURE 6.2
Fit of a haphazardly selected initial parameter vector to the data at four wavelengths.

The sum

$$\sum_{i=2}^{n} p_i = 1,$$

and the p_i are chosen to be a discrete approximation to a beta distribution scaled between a minimum and a maximum activation energy. The beta distribution is chosen because of its flexibility, convenience, and the fact that it includes a uniform distribution that has been shown to be a good approximation to defect distributions in other work (LuValle et al., 1998). This is a very specific form of the model given in Equation 6.2. The model given above has seven adjustable parameters corresponding to the physical phenomena, and then one adjustable parameter for each device (length of fiber) corresponding to the uncertainty in the starting time for the measurement of the anneal of each length of fiber (experimentally there is an unavoidable delay between the time the fiber is placed in the hot oven and the end of the first measurement).

A starting point for the parameter vector is shown in Complement 6.5. The fit at that starting point is given in Figure 6.2, and the final fit in Figure 6.3.

In these figures, the plots labeled fits 1 through 4 show the fits to wavelengths one through four in terms of reduction in loss in decibels per unit

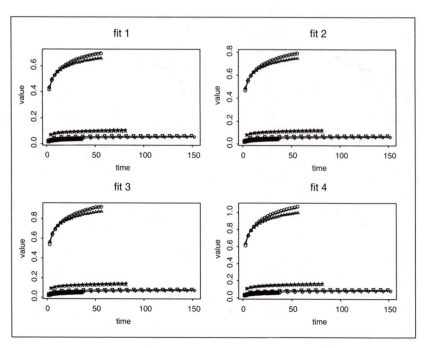

FIGURE 6.3
Fit of optimized parameter vector to the data at four wavelengths.

length (vertical axis) vs. time at the annealing temperature (horizontal axis). The different point styles are from different experimental cells. The fitting process is detailed in Complement 6.5. The parameters for the final fit are

$\nu = 10^{5.73}$ Hz

$E_{a\,min} = 0.33$ eV

$E_{a\,max} = 1.02$ eV

$ypow = 1.15$

$k_1 = 0.006$

$\alpha = 0.652$

$\beta = 1.44$

Data Analysis for Degradation Data

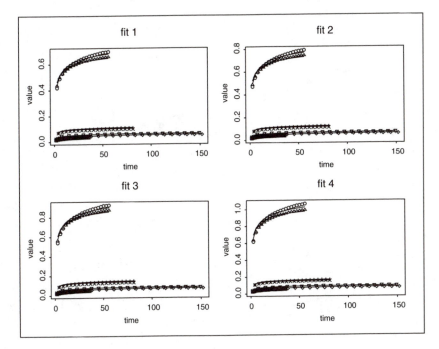

FIGURE 6.4
Fit of optimized parameter vector to the data at four wavelengths fixing $ypow = 1$.

$\Delta t_1 = 5.5$ min, $\Delta t_2 = 4.0$ min

$\Delta t_3 = 5.9$ min, $\Delta t_4 = 4.6$ min

$\Delta t_5 = 4.1$ min, $\Delta t_6 = 1.7$ min

$\Delta t_7 = 2.8$ min, $\Delta t_8 = 2.8$ min

The Δt_i are the estimated times at temperature corresponding to the time between the irradiation and the first measurement of the anneal.

The parameter $ypow$ is rather an odd number, between 1 and 2. This indicates either a very complex nonlinear process or a mixture of first- and higher-order processes governing the formation of lossy defects. To check this, we fit the same model to the data, this time forcing $ypow = 1$. The fit is shown below to the data in Figure 6.4. In Figure 6.5 we show the prediction to the condition given in Table 6.2 for both models. Both predictions are at 99% confidence.

The confidence bounds for the power law model (marked OH and OL) do not contain the observed data, but the confidence bounds for the model power

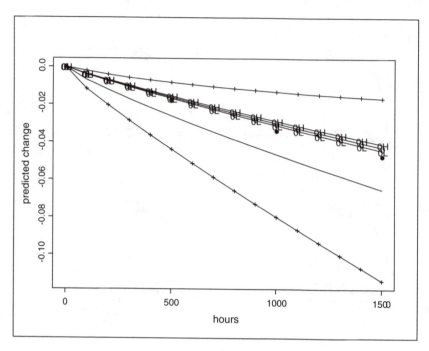

FIGURE 6.5
Extrapolation to 5e−4 dose rate for both models at 95% confidence, plus data.

forced to 1 do. Thus, our tentative conclusion from these data is that the power law model overfits the data, and the simpler model provides a more realistic fit.

The reader is encouraged to continue with the exercises in the complement. There further analysis of this example is carried out, under slightly different assumptions.

6.4 Complement: Background Statistical Theory

The purpose of this section is to provide background in some statistical theory for the analysis of degradation data. The first subsection provides the simple theory for linear regression, from which we build to the rest. References to further reading will be provided in each section.

6.4.1 Linear Regression and Results

In statistics, the linear model has the form: $Y = X\beta + \varepsilon$ (Sheffe, 1959). Here Y is an $n \times 1$ vector of observations, X is a fixed $n \times p$ matrix of predictor variables, β is a $p \times 1$ vector of unknown parameters, and ε is an $n \times 1$ vector of independent and identically distributed measurement errors. Assume that

$E(\varepsilon) = 0$, as otherwise simple reexpression can make it so. To find the least-squares estimates $\hat{\beta}$ of β we start by writing out the sum of squares and differentiating it with respect to the $\hat{\beta}_j$.

$$\|Y - X\hat{\beta}\|^2 = \sum_{i=1}^{n}\left(Y_i - \sum_{j=1}^{p} X_{ij}\hat{\beta}_j\right)^2 = SS$$

$$\frac{dSS}{d\hat{\beta}_j} = -2\sum_{i=1}^{n}\left(Y_i - \sum_{k=1}^{p} X_{ik}\hat{\beta}_k\right)X_{ij} = 0, j = 1,...p \tag{6.10}$$

or

$$\sum_{i=1}^{n}\sum_{k=1}^{p} X_{ij} X_{ik} \hat{\beta}_k = \sum_{i=1}^{n} X_{ij} Y_i, j = 1,...,p$$

The last line can be written more compactly in matrix form:

$$X'X\hat{\beta} = X'Y \tag{6.11}$$

which are called the normal equations. If the columns of X are linearly independent, then the least-squares estimate $\hat{\beta}$ has the form:

$$\hat{\beta} = (X'X)^{-1} X'Y \tag{6.12}$$

An alternative way to think about the least-squares fit is to consider Y as a vector in n-dimensional Euclidean space R^n, and the matrix X as a set of vectors defining a p-dimensional subspace $[X]$ of R^n. Then, the value of $\hat{\beta}$ that minimizes the term $\|Y - X\hat{\beta}\|^2$ is just that value such that $X\hat{\beta}$ is the projection of Y onto $[X]$. Thus, the vector $(Y - X\hat{\beta})$ is orthogonal to $X\hat{\beta}$ and minimizes the squared norm. If the columns of X are not linearly independent, then $\hat{\beta}$ is not unique, and if $(X'X)^-$ is used to denote a generalized inverse of $(X'X)$, then $\hat{\beta} = (X'X)^- X'Y$ is a least-squares estimate. For simplicity, further discussion will be assuming that $\hat{\beta}$ is unique.

Assume the errors, $\varepsilon_i, i = 1,...n$ are independent with identical variance, σ^2, and 0 mean. Then the expected value of $\hat{\beta}$ is β, and the covariance matrix is

$$E\left[(\hat{\beta}-\beta)(\hat{\beta}-\beta)'\right]$$

$$= E\left[\left((X'X)^{-1}X'(Y-(Y-\varepsilon))\right)\left((X'X)^{-1}X'(Y-(Y-\varepsilon))\right)'\right]$$

$$= (X'X)^{-1}X'E(\varepsilon\varepsilon')X(X'X)^{-1} \qquad (6.13)$$

$$= (X'X)^{-1}X'(\sigma^2 I)X(X'X)^{-1}$$

$$= \sigma^2 (X'X)^{-1}$$

By the central limit theorem (Chung, 1974; Serfling, 1980) the distribution of the difference $\sqrt{n}(\hat{\beta}-\beta)$ converges to a Gaussian distribution with mean 0 and covariance matrix $n\sigma^2(X'X)^{-1}$ (note that the elements of $(X'X)^{-1}$ are $O(n^{-1})$ so the covariance matrix stays bounded). If the original errors are Gaussian, then this result is exact, and the least-squares estimate is also the maximum likelihood estimate, which has certain optimal properties, such as achieving the minimum variance among all estimates of β.

6.4.2 Extension to Nonlinear Regression

Unfortunately, very few kinetic models can be expressed in terms of linear regression, so it is necessary to extend the results above to the nonlinear case. In its simplest form, the general nonlinear regression model is

$$Y_i = f(\theta, X_i) + \varepsilon_i \qquad (6.14)$$

Here both θ and X_i may be vectors, Y_i is assumed to be a scalar (a number), and the ε_i are assumed to follow the same simplifying assumptions we have made above. The actual form for the kinds of data we are interested in are often a bit more complex, but we start by looking at results for this kind of form. Under very reasonable assumptions on the form of f and θ, the least-squares estimate $\hat{\theta}$ of θ has been shown to be asymptotically consistent (converge to the true value of θ for large samples) and asymptotically normal (Gaussian) (Jennrich, 1969; Serfling, 1980). Assuming we are close to the true value of θ, the expression in Equation 6.14 can be expanded:

$$Y_i = f(\hat{\theta}, X_i) + \left(f(\theta, X_i) - f(\hat{\theta}, X_i)\right) + \varepsilon_i$$

$$= (\theta - \hat{\theta})\frac{df(\hat{\theta}, X_i)}{d\theta} + f(\hat{\theta}, X_i) + \varepsilon_i + o(\theta - \hat{\theta}) \qquad (6.15)$$

Data Analysis for Degradation Data

where the derivative is a vector derivative. The notation $\delta_n = o(x_n)$ means

$$\frac{\delta_n}{x_n} \xrightarrow[n\to\infty]{} 0$$

Similarly $\delta_n = O(x_n)$ means $0 < (\delta_n/x_n) < C < \infty$ as n increases. The notations o_p and O_p have similar meanings except they are true with probability approaching 1.

Rearranging so that the Y_i are in a single vector, and using appropriate generalizations of the notation, this reduces to

$$\left(Y - f(\hat{\theta}, X)\right) \approx (\theta - \hat{\theta}) \frac{df(\hat{\theta}, X)}{d\theta} + \varepsilon \tag{6.16}$$

which is precisely the linear model formulation. The iterative method based on using Equation 6.16, plus the methods used in linear models, repeatedly to work through a series of estimates of θ is called the Gauss–Newton method. The actual method used in the freeware to do this fitting involves alternating a modification of the Gauss–Newton method based on numerical derivatives, with a modification of the Powell conjugate gradient algorithm (Powell, 1964). This alternation is repeated until a convergence criterion is reached. What is important here is not the fitting algorithm (which we hope will be improved) but the form of Equation 6.16 and the implications for the asymptotic covariance matrix. In particular, the covariance matrix of the normalized difference: $n^{1/2}(\theta - \hat{\theta})$ is simply

$$n\left(\left(\frac{df(\hat{\theta}, X)}{d\theta}\right)'\left(\frac{df(\hat{\theta}, X)}{d\theta}\right)\right)^{-1}$$

Here we assume that the rows of $\left((df(\hat{\theta}, X)/d\theta)\right)$ are simply the derivatives for a single observation Y_i. There are several complications that arise in this situation around both fitting and inference. Bates and Watts (1988) provide a basic reference for nonlinear regression. The data we encounter most often in our accelerated tests are slightly more complicated than the formulation presented here. The next subsection discusses the simple first-order theory used in the freeware to develop confidence intervals for degradation data.

6.4.3 Extension to an Uncertain Starting Time Model

One of the most common stresses used in accelerated testing is temperature. The fastest way to get a device up to temperature is to shove it into a hot oven. However, even when this is done, typically the device takes some time to come to thermal equilibrium, and the first measurement takes time to complete. To model this extra error in degradation measurements on a device, we model the *j*th measurement on device *i* as

$$Y_{ij} = f(\theta, X_i, t_j + \Delta_i) + \varepsilon_{ij} \tag{6.17}$$

where Δ_i is the uncertainty in the time for device *i*, $\vec{Y}_i$ is an n_i-dimensional vector of observations on test unit *i*, θ is a *p*-dimensional vector of physical parameters $\vec{t}_i$ and $\vec{\varepsilon}_i$ are n_i-dimensional vectors of times and measurement errors, respectively, X_i is a matrix describing how stress varies through time, and Δ_i is the actual time offset (unknown) from the recorded times. We assume that *f* is a function of known form. In our case the form of *f* is a first- or zero-order linear kinetic model.

Applying least-squares fitting to estimate the parameters θ and Δ_i in Equation 6.17, when there are several devices (*n*) with device *i* having n_i measurements, we find that the objective function we are trying to minimize has the form (where the $\hat{\theta}$ and $\hat{\Delta}_i$ refer to the parameter estimates):

$$\sum_{i=1}^{n} \left\| Y_i - \hat{Y}_i \right\|^2 = \sum_{i=1}^{n} \left\| \left(f(\theta, X_i, \vec{t}_i + \Delta_i) + \vec{\varepsilon}_i - f(\hat{\theta}, X_i, \vec{t}_i + \hat{\Delta}_i) \right) \right\|^2 \tag{6.18}$$

In accelerated testing, it is possible to bound the Δ_i based on how the experiments were run. To achieve asymptotic consistency (uniqueness) for the estimates of the Δ_i it is necessary to assume a certain nonlinearity of *f* with respect to local displacements in time. Assume that A_0 is the state vector for the material system at time 0, and an observable indexed by λ, Then, for a first-order system of the form:

$$\frac{dA_t}{dt} = K(T)A_t \tag{6.19}$$

and an observable of the form:

$$Y_{it} = c_\lambda A_{it} \tag{6.20}$$

where *i* indexes the device, and c_λ is a vector. The time evolution of the observable has the form: $Y_t = c_\lambda (\exp(K(T)t)) A_0$.

For a given index of the observable, λ, the equation to solve for a constant stress exposure to determine Δ_i is

$$\vec{Y}_i = c_\lambda \left[\left(\exp(K(T)(\vec{t}_i + \Delta_i)) A_0 \right) \right] - c_\lambda \left[\left(\exp(K(T)(\Delta_i)) A_0 \right) \right]$$

or more clearly

$$\vec{Y}_i + c_\lambda\left[\left(\exp(K(T)(\Delta_i))A_0\right)\right] = c_\lambda\left[\left(\exp(K(T)(\vec{t}_i + \Delta_i))A_0\right)\right]$$

Changing any of the parameters results in an affine change to the vector on the left-hand side and either a scale or a nonlinear change to the vector on the right-hand side. Thus, there can be at most one solution to this nonlinear system of equations when the dimension of the observable vector is above the number of free parameters.

Assuming reasonable conditions on the measurement errors, for large values of n and n_i the estimates in Equation 6.18 are asymptotically consistent and normal (Jennrich, 1969). Taking a Taylor expansion of a single difference inside the norm signs:

$$\begin{aligned}
&\left(f(\theta, X_i, \vec{t}_i + \Delta_i) + \vec{\varepsilon}_i - f(\hat{\theta}, X_i, \vec{t}_i + \hat{\Delta}_i)\right) \\
&= \varepsilon_i + \left(f(\theta, X_i, \vec{t}_i + \Delta_i) + \vec{\varepsilon}_i - f(\hat{\theta}, X_i, \vec{t}_i + \Delta_i)\right) \\
&\quad + \left(f(\hat{\theta}, X_i, \vec{t}_i + \Delta_i) + \vec{\varepsilon}_i - f(\hat{\theta}, X_i, \vec{t}_i + \hat{\Delta}_i)\right) \\
&= \varepsilon_i + (\theta - \hat{\theta})\frac{df(\theta, X_i, \vec{t}_i + \Delta_i)}{d\theta} \\
&\quad + (\hat{\Delta}_i - \Delta_i)\frac{df(\theta, X_i, \vec{t}_i + \Delta_i)}{d\vec{t}_i} + O_p\left(\left(n^{-\frac{1}{2}}\right)\left(n_i^{-\frac{1}{2}}\right)\right)
\end{aligned} \tag{6.21}$$

Substituting Expression 6.21 into Expression 6.16, it can be seen that asymptotically for large values of n and n_i, the least-squares estimate of this problem with a shifted timescale is equivalent to a problem with the error vector for the ith device:

$$\vec{\eta}_i + \vec{\varepsilon}_i = \vec{\varepsilon}_i + (\hat{\Delta}_i - \Delta_i)\frac{df(\theta, S_i, \vec{t}_i + \Delta_i)}{d\vec{t}_i} \tag{6.22}$$

The number n_i is typically quite large for each i, so this approximation is assumed to be good from now on. If we define: $\tilde{X} = \left(df(\theta, X, \vec{t} + \bar{\Delta})/d\theta\right)$ where the matrix X and the vectors $\vec{t}$ and $\bar{\Delta}$ are appropriately defined, then the marginal covariance matrix for $(\theta - \hat{\theta})$ has the form:

$$\Sigma_{(\theta-\hat{\theta})} = E\left((\tilde{X}'\tilde{X})^{-1}\tilde{X}'(\varepsilon + \eta)(\varepsilon + \eta)'\tilde{X}(\tilde{X}'\tilde{X})^{-1}\right) \tag{6.23}$$

We assume that measurement error, ε, is independent and identically distributed, while the error term η is asymptotically independent of ε, but internally correlated. Thus, Equation 6.23 reduces to

$$\Sigma_{(\hat{\theta}-\theta)} = \left(\left(\tilde{X}'\tilde{X}\right)^{-1} \tilde{X}' E(\varepsilon'\varepsilon) \tilde{X} \left(\tilde{X}'\tilde{X}\right)^{-1} + \left(\tilde{X}'\tilde{X}\right)^{-1} \tilde{X}' E(\eta'\eta) \tilde{X} \left(\tilde{X}'\tilde{X}\right)^{-1} \right) \quad (6.24)$$

$$= \sigma^2 \left(\tilde{X}'\tilde{X}\right)^{-1} + \left(\tilde{X}'\tilde{X}\right)^{-1} \tilde{X}' \Sigma_\eta \tilde{X} \left(\tilde{X}'\tilde{X}\right)^{-1}$$

where $E(\varepsilon\varepsilon') = \sigma^2 I$, with σ^2 the variance of the measurement error, and $E(\eta\eta') = \Sigma_\eta$, the covariance matrix of the η terms. The form of Σ_η is quite easy to compute up to a constant using Equations 6.19, 6.20, and their solution. In the freeware we match the components of Σ_η to the observed autocorrelation in the data to calculate a scaling factor for estimating Σ_η. We calculate σ^2 by a simple moment's matching to the variance remaining.

In this section we have described both a statistical model of degradation data (Equation 6.17) and our physical model of degradation data (Equations 6.19 and 6.20). We note that, in general, different devices may degrade differently under similar conditions if they have different initial state vectors. However, the average degradation for a set of devices under identical stress conditions under Equations 6.19 and 6.20 is the degradation of a device with the average initial state vector. Thus, under this model with diffuse data we can take k random samples from a population of interest, subject each random sample of n_k devices to identical stress (the k samples going into different stress cells). Then if n_k is large enough, the average path from each cell should behave as if each it is a device with identical (within small error) initial state vectors. If Relation 6.20 were made nonlinear, this would not be true.

6.4.4 Prediction Uncertainty and Asymptotic Relative Efficiency

In addition to calculating uncertainty in the parameter estimates, which can be used to provide confidence bounds and do hypothesis tests about the parameters, it is also important to calculate uncertainty in the predictions. In all cases, the variance of each prediction can be derived as a quadratic form in terms of the covariance matrix of the parameter.

Thus, in the case of degradation data, when we wish to know the variance of the predicted degradation $f(\hat{\theta}, X_{pred}, t_0)$ at a new condition, X_{pred} at a time t_0, we take the vector of derivatives of the prediction with respect the parameters, and calculate

$$\sigma^2_{pred,t_0} = \left(\frac{df(\theta, X_{pred}, t_0)}{d\theta} \right)' \Sigma_{(\hat{\theta}-\theta)} \left(\frac{df(\theta, X_{pred}, t_0)}{d\theta} \right) \quad (6.25)$$

Data Analysis for Degradation Data

Similarly for the failure time data, where predictions are the probabilities of failure at given time, for given stress conditions, the variance of the prediction has the form:

$$\sigma^2_{pred,t_0} = \left(\frac{dG(\phi, X_{pred}, t_0)}{d\phi}\right)' \Sigma_{\hat{\phi}} \left(\frac{dG(\phi, X_{pred}, t_0)}{d\phi}\right) \quad (6.26)$$

Here we are now using the general X_{pred} to denote the environmental conditions for the prediction.

All of the covariance and variance calculations are asymptotic in the sense that these values are correct for very large samples (with appropriate normalization, taken care of in the software).

Once an investigator recognizes that there is uncertainty about statistical estimates, the rather difficult problem occurs that the uncertainty may be so great that it is not yet reasonable to make a decision about what course of action to follow (e.g., the confidence bounds on the probability of failure may be too wide). Alternatively, there may not be enough information to choose which of two (or more) models best represents the phenomena under study.

For this reason, we need some theory to determine how much new experiments may improve the precision of the estimates. In particular, given a model and two experiments to start with, the asymptotic relative efficiency can be calculated for the two experiments. For parameter estimates, with a covariance matrix $\Sigma_{\hat{\theta}_1}$ for the first set of experiments, and $\Sigma_{\hat{\theta}_2}$ for the second, the asymptotic relative efficiency is just the vector of ratios of the diagonal elements of the two matrices. The diagonal elements are just the variances for the individual parameters. Similarly, the ratio

$$ARE_{pred,t_0} = \left(\sigma^2_{pred,t_0,1} / \sigma^2_{pred,t_0,2}\right)$$

is the asymptotic relative efficiency for the prediction. In the latter case, this ratio is the ratio of sample sizes for the two experiments to get identical prediction variance. So if $ARE_{pred,t_0} = 100$, then if n samples are used for experiment 2, $100 \times n$ samples would be required for experiment 1 to obtain the same precision in the estimates.

6.5 Complement: Using the Software to Analyze the Example Data

We begin by creating the base model for the analysis in Section 6.3. In this case, it is simply the "glassy onestep" model found under the "create a

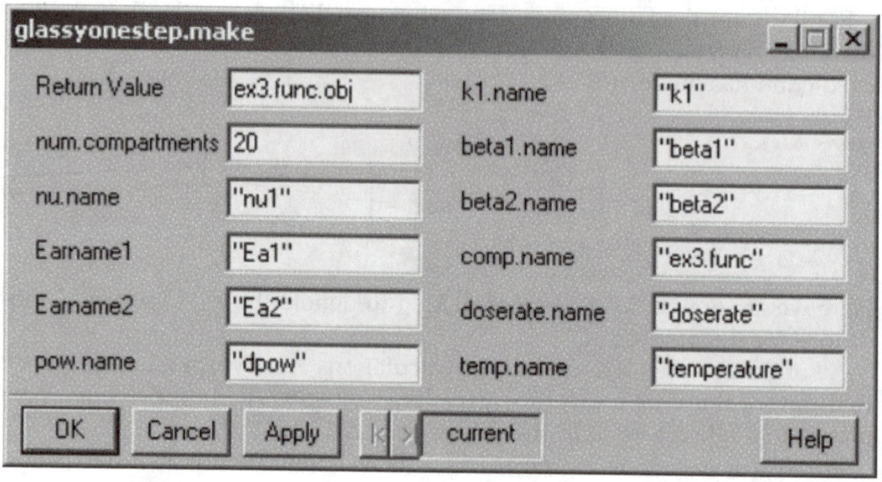

FIGURE 6.6
GUI for creating the kinetic model used to analyze the data in this chapter.

component of the process" submenu under the "kinetic model creation" menu. The GUI is shown in Figure 6.6 filled out to create a 20-compartment model (20 compartments of defects). The output object has the form:

```
> ex3.func.obj
$parameter.vec:
[1] "nu1"     "Ea1"     "Ea2"     "dpow"    "k1"      "beta1" "beta2"

$envir.vec:
[1] "doserate"    "temperature"

$func:
function(parvec, envvec)
{
nu1 <- parvec[1]
Ea1 <- parvec[c(2)]
Ea2 <- parvec[3]
dpow <- parvec[4]
k1 <- parvec[5]
beta1 <- parvec[6]
beta2 <- parvec[7]
doserate <- envvec[1]
temperature <- envvec[2]
Earange <- c(Ea1, Ea2)
Eadistmatbeta(20, doserate, k1, dpow, nu1, Earange,
      beta1, beta2, temperature)
}

$comp.name:
[1] "ex3.func"
```

Data Analysis for Degradation Data 215

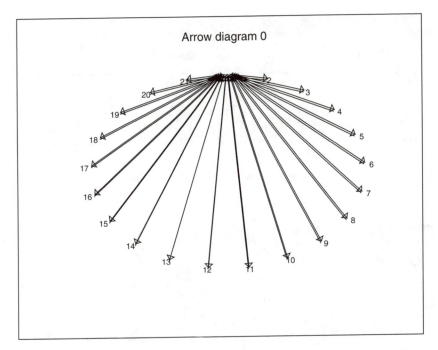

FIGURE 6.7
Arrow diagram for the kinetic model created by the GUI in Figure 6.6.

```
$follow.vec:
 [1] "ex3.func"  "ex3.func"  "ex3.func"  "ex3.func"  "ex3.func"
 [6] "ex3.func"  "ex3.func"  "ex3.func"  "ex3.func"  "ex3.func"
[11] "ex3.func"  "ex3.func"  "ex3.func"  "ex3.func"  "ex3.func"
[16] "ex3.func"  "ex3.func"  "ex3.func"  "ex3.func"  "ex3.func"
[21] "ex3.func"


$workspace.vec:
[1] "Eadistmatbeta"  "ex3.func"
```

The arrow diagram (see Complement 3.2.1.2.1) is shown in Figure 6.7. To join the model and the data, left-click on the "kinetic model analysis" drop-down menu, go to the data analysis branch, and left-click on "construct a degradation model data structure with multiple outputs or latent structure." This brings up the GUI "highlevel.start2." As with the failure time data, there is then a sequence of GUIs brought up that allows us to specify the data structure. These are shown as Figure 6.8 through Figure 6.12.

In the GUI shown in Figure 6.8 we specify four outputs all reflecting the same kinetic structure. Latent structure allows us to have the different outputs reflect different mixtures of the observable part of the state vector. The default for latent structure is a single mixture, with a scaling factor changing the mixture between observables. As it is, we simply specify a positive

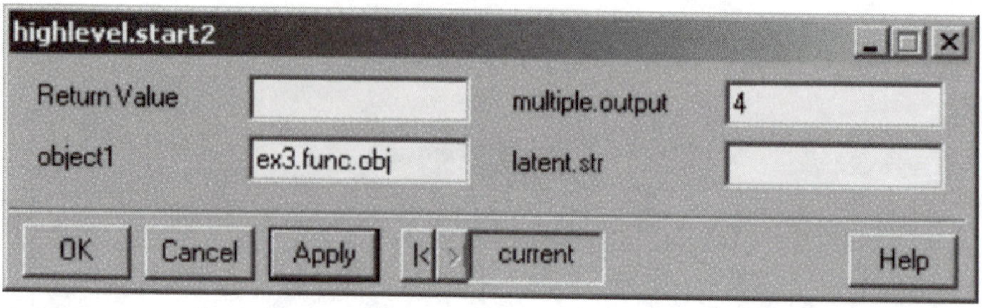

FIGURE 6.8
GUI filled in to start creation of model data object used in the optimization.

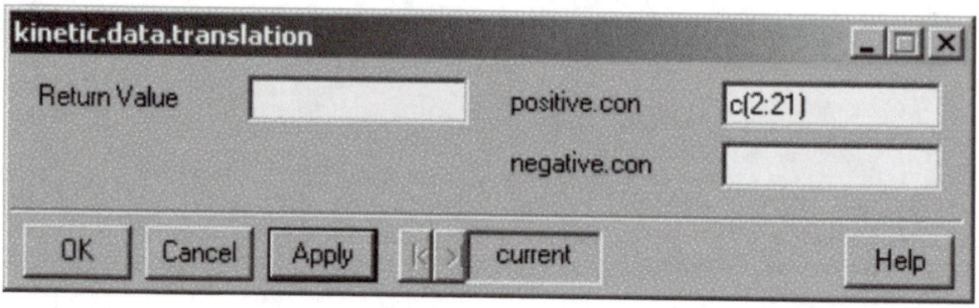

FIGURE 6.9
Second GUI in series used to create model data object.

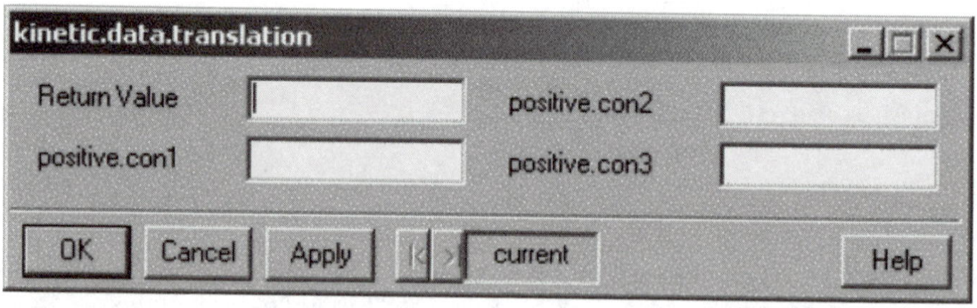

FIGURE 6.10
Alternate to second GUI. This is the GUI that would come up if we wished to represent the data as a complicated linear combination of three latent processes, as opposed to just a sum of states times a single scalar multiplier. This would come if Figure 6.8 had a 3 in the latent process slot.

contribution for the underlying structure as in the GUI in Figure 6.9, reflecting components 2 to 21 of the state vector. Note, in this version of the software, that selecting a set of components to contribute negatively will give an error message and will be ignored in constructing the object.

If a number of latent structures had been selected (say, three) Figure 6.10 would be the GUI that popped up. In that case we would specify components

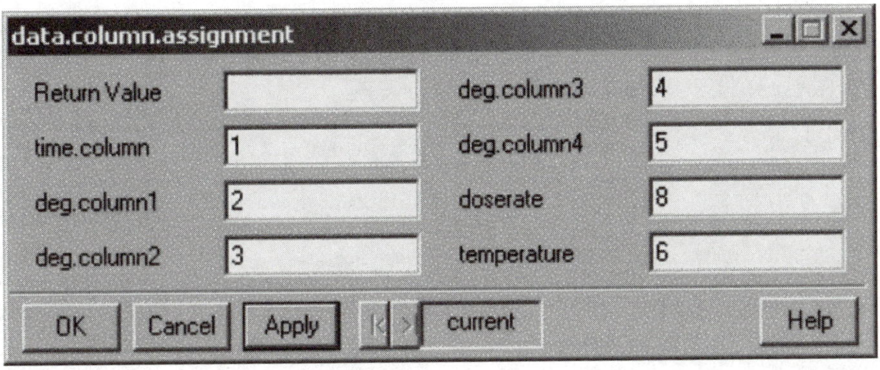

Data Analysis for Degradation Data

FIGURE 6.11
GUI assigning columns of the data matrix to variables specified in the kinetic model object created by the GUI in Figure 6.6. Third GUI in series creating the model data object.

of the state vector for each latent structure that goes into the mixture for each of the (in this case four) observables.

To remember how to fill out the next GUI, specifying the connection between variables for the function and the data structure, we can check by typing in the commands window while the GUI is sitting there. The first five rows of the data matrix from the first section of example3a.dat0 appear as follows:

```
> example3a.dat0[[1]]$mat[1:5,]
          [,1]       [,2]       [,3]       [,4]       [,5]   [,6]
param 0.00000000 0.00000000 0.00000000 0.00000000 0.00000000 110.1
param 0.04333333 0.06794357 0.07403219 0.07943559 0.08451605 109.9
param 0.08638889 0.09326613 0.10342741 0.11798394 0.13096786 110.0
param 0.12944444 0.13189518 0.14596772 0.16314518 0.17556453 110.0
param 0.17277778 0.14741933 0.16181445 0.18443549 0.20790315 109.7
      [,7] [,8] [,9]
param  24   0    1
param  24   0    1
param  24   0    1
param  24   0    1
param  24   0    1
```

The first column is time, the second through fifth columns are degradation measurements, the sixth column is furnace temperature, the seventh is measurement instrument temperature, the eighth is dose rate, and the ninth is a statistical weight corresponding to the precision of the measurement. This is shown in Figure 6.11, where we specify the column assignments in the "data.column.assignment" GUI. Applying this GUI brings up the GUI "highlevel.make2," which allows the final assignment of data analysis parameters shown in Figure 6.12.

This specifies the data set, as with the failure time data, it specifies the number of evenly spaced points along each degradation (or in this case

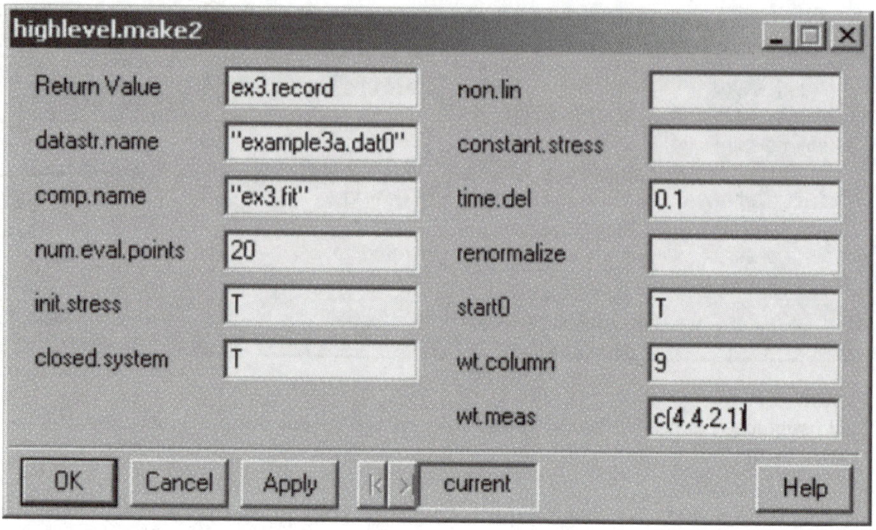

FIGURE 6.12
Final GUI in series creating model data object. Data set is specified, as well as the number of points along the function at which function evaluation takes place. In addition, the following are provided. The time delay (between exposure and start of measurement), weights both for data sets (specified in the data matrices) and for columns of output data (the slot wt.meas), and specification that there is an initial stress exposure before measurement.

annealing) path at which the data will be compared to the model (20). Here the initial stress is in fact very important, as it carries all the information about what dose rate and dose occurred. The kinetic system is closed (first order, not zero order). The nonlinear option is not enabled in this version so leave it on the default (False). The analysis is carried out on constant stress data (this option has not yet been fully implemented either, so again leave on the default). Renormalize means that the data should be divided by the first data point. In some situations this makes sense, particularly when a percent change is the physically appropriate variable. In this case, it is left on the default. However, the data starts at 0 in this data set (loss changes from the first measurement are subtracted out). The weight column for the different data sets is nine, and in this case, the weights on the wavelength measurements are 4, 4, 2, and 1, reflecting the accuracy we believe that each of the four wavelengths provides.

To see how to set up the starting parameter, we find the attributes of the object "ex3.record" we just created, and look at the parameter.vec attribute:

```
> attributes(ex3.record)
$names:
[1] "parameter.vec"  "func"           "comp.name"
[4] "workspace.vec"

> ex3.record$parameter.vec
  [1] "nu1"          "Ea1"          "Ea2"          "dpow"
```

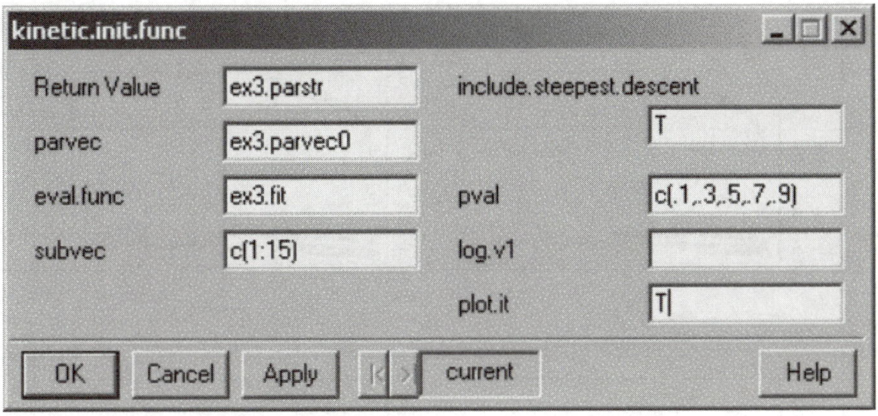

Data Analysis for Degradation Data

[Dialog box: kinetic.init.func]

Return Value	ex3.parstr	include.steepest.descent	T
parvec	ex3.parvec0		
eval.func	ex3.fit	pval	c(.1,.3,.5,.7,.9)
subvec	c(1:15)	log.v1	
		plot.it	T

[OK] [Cancel] [Apply] [< >] [current] [Help]

FIGURE 6.13
Gauss–Newton one-step optimizer set to work from initial value matrix.

```
 [5] "k1"        "beta1"      "beta2"      "time.unc1"
 [9] "time.unc2" "time.unc3"  "time.unc4"  "time.unc5"
[13] "time.unc6" "time.unc7"  "time.unc8"

> ex3.parvec0<-c(10,.1,2,1,.01,1,1,rep(1,8))
```

The time uncertainty is parameterized as before, time*exp(-time.unc), The fit for this roughly random starting point is shown in Figure 6.2, created by typing

```
> ex3.fit(ex3.parvec0,T)$ss
[1] 18.47093
```

in the command window. The number 18.47093 is the squared error. One pass of the Gauss–Newton optimizer, filled out as in Figure 6.13, reduces the sum of squares to approximately 1.14. The graphical output of this GUI is shown in Figure 6.14.

The saved output of the GUI is

```
> ex3.parstr
$theta.del:
 [1] -0.132574287  0.080902883 -0.096840203  0.106998918 (this is the
                                                          direction
 [5]  0.000434911 -0.382246192  0.471112805  0.308810820  that the new
                                                          vector
 [9]  0.019606646  0.363944830 -0.213017981 -0.244042956  is offset)
[13]  0.624628179 -0.363706185  0.505898665

$ss:
[1] 1.137055
```

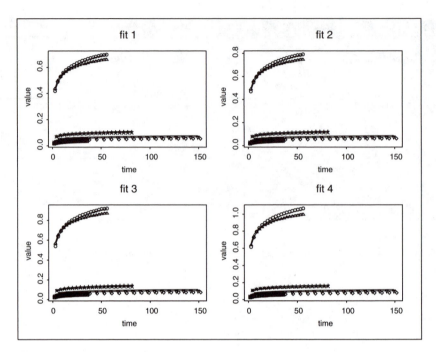

FIGURE 6.14
Graphical view of improved fit after one pass of the Gauss–Newton one-step optimizer.

```
$parvec1:
  [1]  9.86742571 0.18090288 1.90315980 1.10699892 0.01043491
  [6]  0.61775381 1.47111281 1.30881082 1.01960665 1.36394483
 [11]  0.78698202 0.75595704 1.62462818 0.63629381 1.50589867

$dir.find:
 [1] T
```

Optimizing from this point with the conjugate gradient algorithm is accomplished by typing the command:

```
> ex3.parvec1<-ex3.parstr$parvec1
```

then calling up the GUI for the conjugate gradient optimization ("fit the model" under the "data analysis branch" submenu of the "kinetic models analysis" menu), and filling it out as in Figure 6.15. Repeated application (revising ex3.parvec1 to the new parameter estimate each time) slowly lowering sigma0 eventually results in the parameter estimate:

```
> ex3.parvec3
  [1] 5.725652125 0.332938222 1.024181872 1.152154835
  [5] 0.006116768 0.652516470 1.443276454 0.089482748
  [9] 0.402744502 0.023243650 0.255953931 0.391537101
```

Data Analysis for Degradation Data

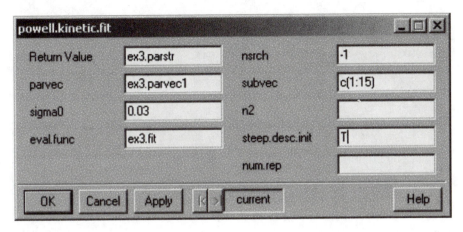

FIGURE 6.15
GUI for conjugate gradient optimizer filled out to continue optimization.

```
[13] 1.262631063 0.776115751 0.754953747
```

The fit to the data is shown in Figure 6.3. The fit is numerically much better than the first pass fit. The command to produce Figure 6.3 is shown below.

```
> ex3.fit(ex3.parvec3,T)$ss
[1] 0.01128129
```

The value of the power variable is rather odd, a number between 1 and 2. To check this we create a starting vector for a reduced model:

```
> ex3.parvec0.1<-ex3.parvec3
> ex3.parvec0.1[4]<-1
```

Now we fit the model, making sure not to alter the power parameter from 1. The GUI to do this is shown in Figure 6.16. Notice that the entry subvec now specifies that the fourth element of the vector is not to be included in the search. The fit after convergence (or convergence as closely as we can tell) has a much larger sum of squares than the fit including the power. The fit is shown in Figure 6.4. The estimate parameter and the sum of squares are shown below:

```
> ex3.parvec0.4
 [1] 0.945368913 0.267755287 1.313666438 1.000000000
 [5] 0.003513462 0.342526827 3.922519583 7.889582959
 [9] 0.285127123 0.234955233 0.025193807 0.046434110
[13] 0.034949955 1.022653804 0.007729164

> ex3.fit(ex3.parvec0.4,T)$ss
[1] 0.2044002
```

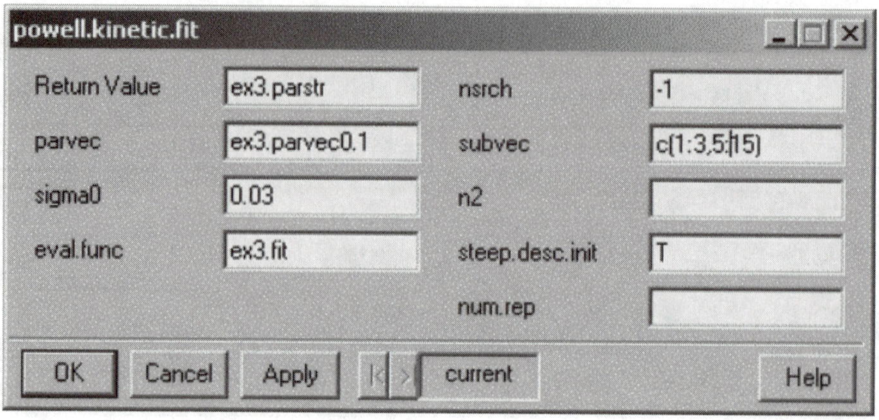

FIGURE 6.16
GUI for conjugate gradient optimizer filled out to work optimization with constraint on power law for the dose rate, confining exponent to 1.

This compares to a sum of squares of 0.011 in the unconstrained search. Unfortunately, it is never easy to tell whether we are overfitting or not. An F test is not appropriate because of the correlation induced by the uncertain starting time. Visually the fit seems reasonable, but consistently underestimates at the highest level.

To make the predictions for degradation data, we left-click the "data analysis branch" submenu under the "kinetic models analysis" menu and left-click "fast parameter covariance estimates, complex models." This brings up the GUI "kinetic.covariance.calc2." We show in Figure 6.17 and Figure 6.18 how we fill this out and apply it twice to make the overlay plot in Figure 6.5. The points are added by a command from the command window. The prediction matrix is created by the command:

```
>ex3.pred.mat.0005<-cbind(rep(100,15),0,0,0,0,25,0,rep(.0005,15))
```

Note the combination slot in the GUI has the vector c(0,1,0,0) in it. This specifies the linear combination of outputs going into the calculation. In particular we are picking out the second wavelength.

Then typing:

```
>data.0005<-cbind(c(500,1000,1500),c(.0181,.0340,.0475))
> points(data.0005[,1],-data.0005[,2],col=3)
```

in the command window creates the plot in Figure 6.5. This plot seems to indicate that the power law may be overfitting the data.

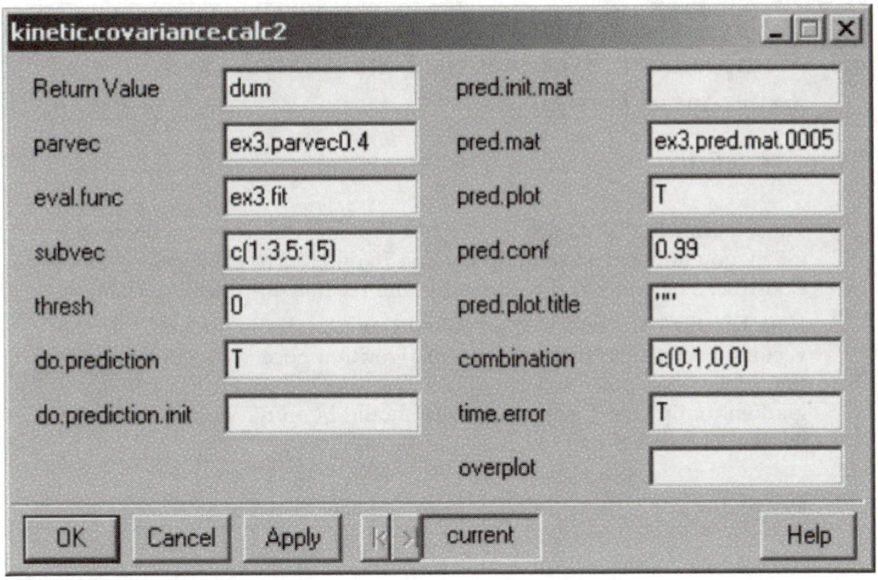

Data Analysis for Degradation Data

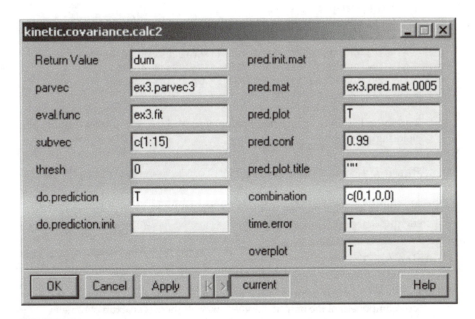

FIGURE 6.17
GUI to calculate and display prediction and 95% confidence bounds for optimized constrained model.

FIGURE 6.18
GUI to calculate and display prediction and 95% confidence bounds for optimized unconstrained model, overlaid on top of the display constructed by applying the GUI in Figure 6.17.

6.6 Complement: Exercises for Data Analysis and Experiment Design

Exercise 6.6.1

The experimentalist comes to you and explains that due to a computer glitch, you may have lost up to the first 18 min (0.3 h) of annealing time. Construct a new model data structure, and see how the inference changes.

Hint: Use the parameter vectors found above as starting points. Run the vectors through the steepest descent algorithm once, only adjusting the time delay parameters. Then run a few times through the conjugate gradient optimizer. Final parameters should be in the neighborhood of the parameters below.

Full model:

```
> ex3.parvec3.2
 [1] 4.778671372 0.327241180 0.992638507 1.135745346
 [5] 0.006194883 0.670418888 1.672288465 1.003917224
 [9] 1.349964960 0.906586091 1.193953877 1.003617911
[13] 1.850602417 1.828787396 1.432303188
```

Restricted model:

```
> ex3.parvec3.1.7
 [1] 5.675638e-001 1.561755e-001 1.731765e+000 1.000000e+000
 [5] 6.235540e-004 1.738636e+000 1.501257e+001 4.463466e+000
 [9] 1.182428e+000 7.017193e-001 6.275673e-001 7.776339e-003
[13] 5.102891e-007 3.612202e+001 1.091778e-001
```

Replotting Figure 6.5 using these parameters and the new model should yield Figure 6.19:

Exercise 6.6.2

Try fitting simpler models to the data, based on one or two activation energies.

Exercise 6.6.3

The covariance matrix from either model fit above is provided under the $cov component of the output. For these models there is an extra parameter corresponding to the linear scaling of the concentration to the loss at the wavelength (given as the $coef component on the output). Unfortunately, it is strongly collinear with the term k_1 in the model. The covariance

Data Analysis for Degradation Data

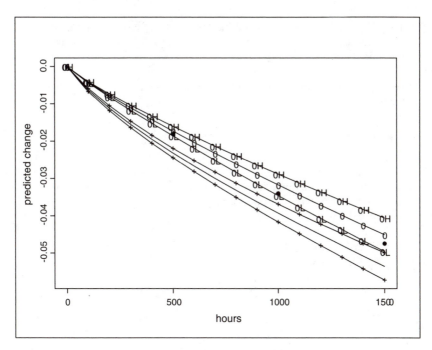

FIGURE 6.19
Display of confidence bounds and raw data for models rebuilt according to the results of Exercise 6.6.1.

matrix should show this. Access the covariance matrix for each of the fits above, and check how large the variance is for k_1 and for the last term in the covariance estimate. Based on the Equations 6.3, 6.8, and 6.9 and what they mean, perform a dimensional analysis to show why these two estimates are so difficult to separate, and what combination of them would be more stable.

Exercise 6.6.4

Build a model assuming two separate classes of defects, so two separate defect creation coefficients, and two separate distributions of activation energies. Fit the model to the data.

Exercise 6.6.5

It is possible to examine the effects of reducing the experiment design by creating new data structures with deleted components. For example, to delete components 1 to 4 (experiments 1 to 4) in this data structure, we can use the command:

```
> example3b.dat0<-example3a.dat0[-c(1:4)]
```

Try deleting the following combinations and see what happens to the estimates and their covariances: 1 and 2, 3 and 4, 1 to 4, 5 to 8, based on the models developed in Exercise 6.6.1.

Hint: Each new data structure involves recreating a model data structure for fitting.

Exercise 6.6.6

There are several kinetic models involving higher-order kinetics that give rise to apparent fractional powers. Give two, one that gives rise exactly to a fractional power and one that gives rise to the fractional power as an approximation.

Hint: Consider the system of equations:

$$\frac{dA}{dt} = k_2 D^6 - k_1 \gamma A$$

$$\frac{dD}{dt} = k_1 \gamma A - k_2 D^6 - k_3 \gamma D$$

$$\frac{dL}{dt} = k_3 \gamma D$$

where D represents a short-lived defect, γ is dose rate, A is the native glass, and L is the loss-causing defect. Solve for the dependence of the generation of the loss-causing defect on γ.

References

Augis, J.A., Denure, D.G., LuValle, M.J., Mitchell, J.P., Pinnel, M.R., and Welsher, T.L. (1989) A humidity threshold for conductive anodic filaments in epoxy glass printed wiring boards, in *Proceedings of the 3rd International SAMPE Electronics Conference*, Lockerby, S.C. et al., Eds., SAMPE, Covina, CA, 1023–1030.

Bates, D.M. and Watts, D.G. (1988) *Nonlinear Regression Analysis and Its Applications*, Wiley, New York.

Benson S.W. (1960) *The Foundations of Chemical Kinetics*, McGraw-Hill, New York.

Benson, S.W. (1976) *Thermochemical Kinetics*, Wiley Interscience, New York.

Braunauer, S., Emmet, P.H., and Teller, E. (1938) Adsorptions of gases in multimolecular layers, *Journal of the American Chemical Society*, 60, 309–319.

Breiman, L. Freidman, J.H., Olshen, R.A., and Stone, C.J. (1984) *Classification and Regression Trees*, Wadsworth International Group, Belmont, CA.

Chernoff, H. (1972) *Sequential Analysis and Optimal Design*, SIAM, Philadelphia.

Chung, K. (1974) *A Course in Probability Theory*, Academic Press, New York.

Cramer, H. (1946) *Mathematical Methods of Statistics*, Princeton University Press, Princeton, NJ.

Crank, M. (1975) *The Mathematics of Diffusion*, Oxford University Press, Oxford.

DeGroot, M.H. (1970) *Optimal Statistical Decisions*, McGraw-Hill, New York.

Efron, B. (1982) *The Jackknife, the Bootstrap, and Other Resampling Plans*, Society for Industrial and Applied Mathematics, Philadelphia.

Elfving, G. (1952) Optimal allocation in linear regression theory, *Annals of Mathematical Statistics*, 23, 255–262.

Erdogan, T., Mizrahi, B., Lemaire, P.J., and Monroe, D. (1994) Decay of ultraviolet induced fiber Bragg gratings, *Journal of Applied Physics*, 76, 73–80.

Ferguson, T.S. (1967) *Mathematical Statistics: A Decision Theoretic Approach*, Academic Press, New York.

Feynman, R.P., Leighton, R.B, and Sands, M. (1963) *The Feynman Lectures on Physics*, Addison-Wesley, Reading, MA.

Glasstone, S., Laidler, K., and Erying, H. (1941) *The Theory of Rate Processes*, McGraw-Hill, New York.

Gong, G. and Sameniego, F. (1981) Pseudo maximum likelihood estimation: theory and applications, *Annals of Statistics*, 9(4), 861–869.

Griscom, D.L., Gingerich, M.E., and Friebele, E.J. (1991) Radiation induced defects in glasses: origin of the power-law dependence on concentration of dose, *Physical Review Letters*, 71, 1019–1022,

Hobbs, G. (2000) *Accelerated Reliability Engineering: HALT and HASS*, Wiley, New York.

Holcomb, D. (2001) Personal communication.

Jennrich, R. (1969) Asymptotic properties of nonlinear least squares estimation, *Annals of Mathematical Statistics*, 40, 633–643.

Kabfleisch, J.D. and Prentice, R.L., (1980), *The Statistical Analysis of Failure Time Data,* Wiley, New York.

Kittel, C. and Kroemer, H. (1980) *Thermal Physics,* W.H. Freeman, San Francisco.

Klinger, D.J. (1991) Humidity acceleration factor for plastic packaged electronic devices, *Quality and Reliability Engineering International,* 7, 365–370.

Klinger, D.J. (1992) Failure time and rate constant of degradation: an argument for the inverse relationship, *Microelectronics and Reliability,* 32, 987–994.

Krauss, A.S. and Erying, H. (1975) *Deformation Kinetics,* Wiley, New York.

Lawless, J. (1982) *Statistical Models and Methods for Lifetime Data,* Wiley, New York.

Levine, I.N., *Physical Chemistry,* McGraw Hill, 1988.

Lindley, D.V. (1972) *Bayesian Statistics, a Review,* CBSM-NSF, SIAM, Philadelphia.

LuValle, M. (1999) An approximate kinetic theory for accelerated testing, *IIE Transactions on Quality and Reliability Engineering,* 31(12), 1147–1156.

LuValle, M. (2000) A theoretical framework for accelerated testing, in *Recent Advances in Reliability Theory,* Limion, N. and Nikulin, M., Eds., 419–433, Birkhauser, Boston.

LuValle, M. (2001) A useful representation for kinetic modeling in accelerated testing, in *Proceedings of the Section on Physical and Engineering Sciences of the American Statistical Association,* [CD ROM] American Statistical Association, Alexandria, VA.

LuValle, M. and Hines, L.L. (1992) Using step stress to explore the kinetics of failure, *Quality and Reliability Engineering International,* 8, 361–369.

LuValle, M.J. and Mitchell, J.P. (1987) The design and interpretation of accelerated life tests for printed circuit products, *Proceeding of the IEPS,* 1, 444–461.

LuValle, M., Welsher, T., and Mitchell, J.P. (1986) A new approach to the extrapolation of accelerated life test data, in *Proceedings of the 5th International Conference on Reliability and Maintainability,* Biarritz, France, 630–635.

LuValle, M., Welsher, T., and Svoboda, K. (1988) Acceleration transforms and statistical kinetic models, *Journal of Statistical Physics,* 52, 311–320.

LuValle, M. Copeland, L.R., Kannan, S., Judkins, J., and Lemaire, P. (1998) A strategy for extrapolation in accelerated testing, *Bell Labs Technical Journal,* 3(3), 139–147.

LuValle, M., Brown, G, Lefevre, B., Reith, L., Throm, R. (2000) Acceptance testing for the pistoning failure mode in fiber optic connectors, *Proceedings of the SPIE,* 4215, 168–182.

LuValle, M., Mrotek, J., Copeland, L., LeFevre, B., Brown, G., and Throm, R. (2002) Integrating computational models with experimental data in reliability, *SPIE,* 4639, Matthewson, M. and Kurkjian, C., Eds., 52–63.

Manson, S.S. (1981) *Thermal Stress and Low Cycle Fatigue,* Robert Krieger, Malabar, FL.

Matsuoka, S. (1992) *Relaxation Phenomena in Polymers,* Hanser, Munich.

Meeker, W. and Escobar, L. (1998) *Statistical Methods for Reliability Data,* Wiley, New York.

Meeker, W.Q. and LuValle, M. (1995) An accelerated life test model based on reliability kinetics, *Technometrics,* 37(2), 133–148.

Miner, J. (1945) Cumulative damage in fatigue, *Journal of Applied Mechanics,* 12, A159–A165.

Nelson, W. (1982) *Applied Life Data Analysis,* Wiley, New York.

Nelson, W. (1990) *Accelerated Testing: Statistical Models, Test Plans, and Data Analysis,* Wiley, New York.

Oppenheim, I, Shuler, K., and Weiss, G. (1977) *Stochastic Processes in Chemical Physics: The Master Equation,* MIT Press, Cambridge, MA.

Palmer, R.G., Stein, D.L., Abrahams, E., and Anderson, P.W. (1984) Models of hierarchically constrained dynamics for glassy relaxation, *Physical Review Letters*, 53, 958–961.

Pecht, M.G. (1994) *Integrated Circuit, Hybrid, and Multichip Module Package Design Guidelines: A Focus on Reliability*, Wiley, New York.

Pecht, M.G. and Nash F.R. (1994) Predicting the reliability of electronic equipment, *Proceedings of the IEEE*, 82(7), 992–1004.

Pecht, M.G., Radojcic, R., and Rao, G. (1999) *Guidebook for Managing Silicon Chip Reliability*, CRC Press, Boca Raton, FL

Powell, M.J.D. (1964) An efficient method for finding the minimum of a function of several variables without calculating derivatives, *Computer Journal*, 7, 155–162.

Primak, W. (1955) The kinetics of processes distributed in activation energy, *Physical Review*, 100(6), 1677–1689.

Roussas, G., *A first course in Mathematical Statistics*, Addison Wesley, 1973.

Scher, H., Schlesinger, M.F., and Bendler, J.T. (1991) Time scale invariance in transport and relaxation, *Physics Today*, 44(1), 26–34.

Serfling, R. (1980) *Approximation Theorems of Mathematical Statistics*, Wiley, New York.

Sheffe, H. (1959) *The Analysis of Variance*, Wiley, New York.

Yost, F.G., Amos, D.E, and Romig, A.D., Jr. (1989) Stress driven diffusive voiding of aluminum conductor lines, in *Proceedings of the International Reliability Physics Symposium*, 193–201, IEEE, Piscataway.

Zwolinsky, B.T. and Erying, H. (1947) The non-equilibrium theory of absolute rates of reaction, *Journal of the American Chemical Society*, 69, 2702–2707.

Appendix: Installing the Software

The software is currently a set of Splus® dump files. They can be downloaded from www.crcpress.com/e_products/downloads/. They have been checked on Splus 2000® and Splus 6.1®. To load the freeware, assuming Splus is running and the software has been downloaded from the Web site:
Copy all files ending ".pck," into your user directory in Splus
Open Splus, and use the commands:

- source("demarcbook.pck")
- source("DES.pck")
- source("EPmap.pck")
- source("ft.pck")
- source("kinetics1.pck")
- source("kinetics2.pck")
- source("example.pck")

This loads the necessary software and constants into Splus.
To install the menus, use the commands:

- create.menu.bookdemarc()
- create.menu.kinetics1()
- create.menu.kinetics2()
- create.menu.ft()
- create.menu.EvanescentProcess()

Finally, bring up the object explorer in Splus (if it is not shown, click on the button on the toolbar that looks like a three-dimensional representation of a cardboard box with an open top, and a circle, triangle, and square above it). Right click on the box to the left side of the object explorer (where the directories are listed) and left click on "insert folder" in the pop-up menu. Name the new folder "functions" and right click on it. Left click on "folder" in the resulting pop-up menu and then left click on the "functions" box in the resulting pop-up menu and finally left click on the apply button.

When you leave Splus after this, you should be prompted if you want to save the object explorer as default. You should indicate that you do.

Index

A

Accelerated failure time model, 138–139
Accelerated testing
 concepts of, 1–2
 data types in, 135
 definition of, 1
 description of, 25
 limitations of, 115–116
 purposes of, 25
 summary of, 115–116
 temperature effects, 209
Acceleration transform, 167
Acceptance tests
 definition of, 26
 mechanical cycling induced by thermal cycling, 41–43
 for purely thermal process, 34–37
 for temperature/humidity models, 38–40
Analytical demarcation maps, 26–27
A priori failure, 4, 26
Arrhenius law, 11–13
Arrhenius model, 77, 101
Arrhenius plot, 7
Arrhenius premultiplier, 113
Asymptotic relative efficiency, 212–213

B

Band gap effect, 77
Bayesian inference, 19
Bernoulli random variable, 17, 51
Beta binomial interpretation of 0 failures, 50–51
Binomial distribution
 description of, 17–18
 inference for, 18–20
Boltzmann distribution, 12
"Breaking strength" data, 135

C

Central limit theorem, 160–161, 208
Chemical rate constant, 17
Classical confidence interval, 18–19
Classical statistical approach, 2–3
Coffin–Manson demarcation maps, 42–43
Coffin–Manson model, 40
Competing reaction operation, 81–85
Computational demarcation maps
 definition of, 27
 examples of, 44–49, 92–96
Conductive anodic filament failure mode, 111–112, 197
Confidence bounds, 158
Confounding, 155
Covariance matrix, 211–212
Creep, 73–74

D

Data
 "breaking strength," 135
 degradation, see Degradation data
 failure time, 135
Degradation, 5
Degradation data
 data analysis for
  exercises for, 224–226
  models, 197–200
  motivation, 197–200
 description of, 135
 example
  background for, 200–201
  data analysis for, 201–206
  exercises for, 224–226
  Splus software for analyzing, 213–223
 statistical theory for
  asymptotic relative efficiency, 212–213
  description of, 206
  linear regression, 206–208
  nonlinear regression, 208–209
  prediction uncertainty, 212–213
  uncertain starting time model, 209–212
Demarcation approximation
 definition of, 27–28
 thermal, 28
 time–temperature exposure, 30

Demarcation array calculator, 125–127
Demarcation maps and mapping
    analytical, 26–27
    classification of, 26–27
    Coffin–Manson, 42–43
    computational
        definition of, 27
        examples of, 44–49, 92–96
    description of, 9
    design of, 34–37
    evanescent process, 116
    extrapolation theorem, 51–52
    mechanical cycling, 61–63
    for multiple experiments, 53–55
    thermal
        description of, 27–34
        using Splus software, 56–58
Discrete diffusion, of kinetic model, 71–73

E

Euclidean space, 207
Evanescent process mapping
    demarcation maps, 116
    description of, 26
    example of, 111–115
    graphical user interface, 116–133
    model enumeration, 102–104
    model neighborhoods
        description of, 98–101
        identification of, 105–111
    overview of, 97–98
    risk orthogonality, 102
    theory integration, 104
Experiments
    demarcation maps for, 53–55
    step stress, 26
Extrapolation theorem, 51–52

F

Failure
    a priori, 4, 26
    beta binomial interpretation of zero failures, 50–51
    physics of, 3
    statistical analysis of, 17
Failure time data
    data analysis for
        data set, 135–146
        real data set, 148–158
    definition of, 135
    example of
        data set for, 148–158
        printed wiring-board data, 171–174
    exercises for, 190–196
    graphical user interface for, 175–189
    Kaplan–Meier estimate, 171
    kinetics estimation from, 166–169
    physical sense added to model, 146–148
    summary of, 158–159
First-order kinetics
    description of, 14–17
    mathematical equivalent representations of, 98
Fluence, 83

G

Gap effect, 77
Gaussian distribution, 169
Gaussian random vector, 163
Gauss–Newton algorithm, 69
Gauss–Newton fit, 185
Gauss–Newton optimizer, 186
Graphical user interface
    degradation data analysis, 213–223
    evanescent process mapping, 116–133
    failure time data, 175–189
    for kinetic models, see Kinetic models

H

HALT, see Highly accelerated life testing
HASS, see Highly accelerated stress screens
Highly accelerated life testing, 3
Highly accelerated stress screens, 3

K

Kaplan–Meier estimate, 171
Kinetic models
    creep, 73–74
    description of, 65, 100
    discrete diffusion, 71–73
    failure time data for estimating, 166–169
    glassy one step, 79–81
    interface for building, 65–70
    one step
        glassy, 79–81
        process of, 75–76
        with second variable, 77
        with second variable and stress, 78–79
        with stress, 77–78
    overview of, 70

Index

stress voiding, 74–75
submodel combining operations
   competing reactions, 81–85
   connection of internal states, 90–92
   mixing reactions, 85–86
   overview of, 81
   rejoining reactions, 87–89
   reversible reactions, 86–87
   sequential reactions, 89–90
Krauss and Erying model, 44

L

Law of large numbers, 160
Least-squares estimate, 207
Least-squares fit, 210
Linear models theory, 65
Linear regression, 206–208
Lognormal plot, 151–152

M

Maps, see Demarcation maps and mapping; Evanescent process mapping
Master equation, 159
Maximum likelihood estimates
   central limit theorem, 160–161
   description of, 140, 142, 159–160
   law of large numbers, 160
   proof of consistency of, 161–162
   pseudo-, 155–156, 169–170, 183, 187
   Splus source code for, 164–166
Maximum likelihood estimator, 162–164
Mechanical cycling
   acceptance testing for, 4143
   demarcation maps for, 61–63
   models of, 40–41
Miner's rule, 41
Mission-critical reliability, 1
Mixing reaction operation, 85–86

N

Nonlinear regression, 208–209
Nonparametric survival curve, 151–152
Nuisance parameters, 183
NULL model, 105

O

One-step process, for kinetic models
   glassy, 79–81
   process of, 75–76

with second variable, 77
with second variable and stress, 78–79
with stress, 77–78

P

Physical statistics
   description of, 3
   overview of, 3–4
Powell algorithm, 69
Powell conjugate gradient algorithm, 209
Power law model, 202, 205
Prediction uncertainty, 212–213
Probability integral transform, 192
Proportional hazards model, 138–139
Pseudo activation energy, 7
Pseudo-maximum likelihood estimation, 155–156, 169–170, 187

R

Reaction rate theory, 6
Rejoining reaction operations, 87–89
Relative humidity
   laws, 13
   measurement of, 149
Reliability
   accelerated testing for determining, see Accelerated testing
   binomial distribution's role in, 17–18
   classical predication approach to, 2
   criteria for, 5
   definition of, 1
   information about, 25
   mission-critical, 1
Reversible graphical user interface, 181
Reversible reaction operation, 86–87

S

Safety tests, 26
Saturation vapor pressure, 13
Scaling time, demarcation approximation used for, 29
Sequential reaction operations, 89–90
Splus software
   degradation data analysis using, 213–223
   description of, 10–11
   graphical user interface for, 56
   how to use, 55–56
   matrix exponentiation in, 20–23
   maximum likelihood estimates analysis, 164–166

temperature/humidity demarcation maps
  using, 58–61
  thermal demarcation maps using, 56–58
Statistical kinetics, 159
Statistical theory
  asymptotic relative efficiency, 212–213
  description of, 206
  linear regression, 206–208
  nonlinear regression, 208–209
  prediction uncertainty, 212–213
  uncertain starting time model, 209–212
Step stress experiments, 26
Stochastic model, 139–140
Stress voiding, 74–75

T

Temperature/humidity models
  acceptance tests for, 38–40
  demarcation maps, using Splus software, 58–61
  simple, 37
Thermal cycling, mechanical cycling induced by, 41–43

Thermal demarcation approximation, 28
Thermal demarcation maps
  description of, 27–34
  using Splus software, 56–58
Thermal process, acceptance tests for, 34–37

U

Unimolecular reaction, 11–12
Unobservable censoring model, 140, 142–143, 145
Unobservable masking model, 100

V

Vapor pressure, 13–14

W

Weibull distribution, 137, 139, 167–168, 192
Weibull plot, 151–152